Nature, Culture, and Society

Life is currently one of the most active zones of politics and economic production, as biological material is increasingly the subject of engineering, banking, reproduction, and exchange. These developments represent some of the most challenging issues facing humanity in the twenty-first century and call for new forms of engagement – and new anthropologies of life.

Reflecting upon the changing human condition, Palsson addresses various conflated zones of life at particular times and scales, from the genome to the human body and the global environment. Using a "biosocial" perspective, he argues, will help us to capture the hybrid nature of life, enhancing our sensitivity to differences and similarities in hierarchies, the reproduction of bio-objects and the exchanges between humans, other species, and the environment.

Engaging with topical issues on the public agenda, from personal genomics to human–animal relations to the global environment, the book sets out a compelling case for meaningful change.

GISLI PALSSON is Professor of Anthropology at the University of Iceland and Visiting Professor at King's College, London.

Nature, Culture, and Society

Anthropological Perspectives on Life

Gisli Palsson

CAMBRIDGE
UNIVERSITY PRESS

CAMBRIDGE
UNIVERSITY PRESS

University Printing House, Cambridge CB2 8BS, United Kingdom

Cambridge University Press is part of the University of Cambridge.

It furthers the University's mission by disseminating knowledge in the pursuit of education, learning and research at the highest international levels of excellence.

www.cambridge.org
Information on this title: www.cambridge.org/9781107085848

First published 2016

A catalogue record for this publication is available from the British Library

Library of Congress Cataloguing in Publication data
Gisli Palsson, 1949–
Nature, culture and society : anthropological perspectives on life / Gisli Palsson.
pages cm
Includes bibliographical references and index.
ISBN 978-1-107-08584-8 (Hardback)
1. Physical anthropology. 2. Sociobiology. 3. Life sciences. I. Title.
GN60.G57 2015
599.9–dc23
2015016633

ISBN 978-1-107-08584-8 Hardback

Contents

Preface

This book brings together a collection of essays that I have written for different books and journals over the past two decades or so. They have been selected for the sake of theoretical and empirical coherence; all of them address some aspects of what I call "anthropologies of life," focusing on genomics, bodies, the environment, and their relations. Some of them are new (Chapters 6 and 12), others have been written at different moments for different agendas. Written in 1994, the oldest essay ("Enskilment at Sea"; Chapter 9) may seem an oddity. It did, however, represent a significant and somewhat surprising shift in my writing and perspective and, moreover, it resonates with much of what I wrote later, providing a link between bodies and environments, practices and politics. I have arranged the essays in three sections to provide a general flow from the molecular basis of life to the body and the global scene. The Introduction and Afterword highlight the context and the main issues covered, situating them in anthropological discussions and in the current "biosocial turn." Assembling the essays, I have edited each of them, sometimes heavily, to connect, update, and avoid repetition. Those who are interested in details that I have omitted should consult "Sources and Acknowledgments" at the end of the book where the relevant original publications are listed.

My work on both the book and the series of projects on which it is based would not have been possible without the friendship and the support I have enjoyed over the years, financial and intellectual. This work has been generously funded by the Icelandic Centre for Research, the University of Iceland, the Nordic Environmental Research Programme, and the Nordic Committee for Social Science Research (NOS-S). Also, invitations to a number of universities and research institutes have been highly useful, alerting me to new avenues and developments. Furthermore, participation in several workshops and collaborative research projects has been personally rewarding and intellectually invigorating.

I thank publishers and editors who have granted permissions to reproduce material earlier published. Also, I am grateful to numerous colleagues, students, friends, journal editors, workshop discussants, assistants, administrators, and informants (if such a term is still permitted) – all of whom have kindly shared with me their thoughts on important issues or taken time to read my drafts at one stage or another, offering comments, correcting my course, and generating further work: David Akin, Emily Angulalik, Debbora Battaglia, Aletta Biersack, Carole H. Browner, James Carrier, Kim Crockatt, E. Paul Durrenberger, Sigríður Sunna Ebenersersdóttir, Yulia Egorova, César Enrique Giraldo Herrera, Herbert Gottweis, James Greenberg, Steven Gudeman, Sigurður Örn Guðbjörnsson, Akhil Gupta, Kristín Erla Harðardóttir, Anna Harris, Kirsten Hastrup, Agnar Helgason, Tim Ingold, Mark Jackson, Susan Kelly, Marianne Lien, Susan Lindee, Margaret Lock, Jonathan Marks, Emily Martin, Mette Nordal Svendsen, Benjamin Orlove, Thomas Osborne, Barbara Prainsack, Tanya Richardson, Laura Rival, Silke Schicktanz, Ingrid Schneider, Henning Sloth Pedersen, Marianne Sommer, Sverker Sörlin, Halldór Stefánsson, Bronislaw Szersynski, Mary Ellen Thomas, Martina Tyrrell, Ricardo Ventura Santos, Matthew Welpton, Gabriele Werner-Felmayer, Gisa Weszkalnys, Sally Wyatt, Þórólfur Þórlindsson, and Höskuldur Þráinsson. No doubt I am missing some of the names in this long-term project. Some have to be omitted; a number of anonymous reviewers have provided extensive help.

Cambridge University Press staff and their reviewers embraced the book concept from the start. I thank all of them, in particular Executive Publisher Andrew Winnard who took the initiative and helped to structure the product and followed it through. The projects on which the essays are based would not have been possible without the support and patience of my family – especially my wife, Guðný. During the final construction of the book, I have to add, our grandchildren – Gísli Þór, Jón Bjarni, and Saga Rós – have been a continuous source of delight, contributing, in their own fashion, new perspectives and insights on anthropologies of life.

GISLI PALSSON, REYKJAVIK

1 Introduction

"Incidentally," Paul Rabinow (1996: 99) remarked in his *Essays on the Anthropology of Reason*, "there is no entry for 'life' in *Keywords*." He was referring to the first edition of Raymond Williams's heavily cited reference work, first published in 1976. "Life," in other words, wasn't of significance in late twentieth-century discourse. How could that possibly be the case? The "omission," if that is the right term, in a work that has proved to be a valuable guide to modern thought, at least Euro-American discourse, is remarkable. Could it be that "life" had not yet emerged as an epistemic object? In a sense, the fascination with *some* concept of life is cross-cultural and panhuman, dating back to early myths at the dawn of humanity. And, after all, the European Enlightenment sparked immense interest in "life," "nature," and related themes. A whole army of Enlightenment naturalists explored and theorized the history of the planet, the evolution of organisms, and different forms of life (see, for instance, Rudwick 2014). Life was intensively scrutinized through the modernist gaze of "disciplines," each with its own intellectual terrain, its schools, chairs, and disciples.

If the Enlightenment marked the birth of biophilia (literally, "the love of life"), the *absence* of "life" from Williams's *Keywords* signified its death, a mysterious disappearance. Originally used, it seems, by Eric Fromm in the early 1960s (Kahn 2011: 11–26), the term "biophilia" has been developed by several scholars for related purposes. During the last century, especially the second half, a new surge of interest in "life" developed. Schrödinger's book *What Is Life?*, originally published in 1944 (Schrödinger 2006), launched a wave of inquiries into "life itself" in the natural sciences, the humanities, and the social sciences (see, for instance, Jonas 2001 and Dupré 2012). This has generated new kinds of concerns and new epistemic spaces (Canguilhem 2008, Rheinberger 2010). Popular culture, too, is saturated with references to life. John Lennon wrote that when he went to school he was once asked what he wanted to be when he grew up. "Happy," he wrote down. His teachers told him he didn't understand the assignment, but Lennon told them they didn't understand life. It seems safe to say that life has recaptured the imagination of the public, environmentalists, and academics across the disciplinary terrain.

Anthropology (following Kohn 2007, Kirksey and Helmreich 2010, and some others) can usefully be expanded and redefined as the study of more than one species, as the "anthropology of life": "to encourage the practice of a kind of anthropology that situates all-too-human worlds within a larger series of processes and relationships that exceed the human" (Kohn 2007: 6). In this perspective, the study of humans is inseparable from the study of other organisms. Many ethnographies already support such an argument, giving voice to those who reject categorical distinctions between humans and other beings that characterize modernist discourse. I take the reference to "anthropologies of life," in the *plural*, to mean several things, including multi-species ethnography, comparisons of ethnobiologies (the endless variety of folk accounts of species differences and relations), anthropological inquiries into "life itself" (sometimes redefining and refashioning *bios* as we know it), and the ways in which anthropologists "dissolve" the human into "life," biological and social. While "social life" and "life itself" were separated by Enlightenment thinkers, they now seem to be recombined under the banner of the "biosocial turn" in several fields (Meloni 2014), including anthropology.

This introduction outlines broad trends of anthropologies of life, the theoretical and empirical terrain covered by the book, some of the literature it draws upon, the key concepts developed, and the themes connecting different chapters. All of the chapters address some aspect of life – problematizing "life itself," documenting and reflecting on human entanglements and the changing human condition, while engaging with the complexities of genomes, bodies, and biospheres. To facilitate understanding of such a broad terrain, I suggest a general theoretical perspective that centers on biosocial relations. In the plural, however, anthropologies of life are not restricted to multiple ethnobiologies, the endless variety of folk accounts of species differences and relations. Also, and more importantly, the term may be taken as inquiries into life itself, the processes and metaphors through which people distinguish (or refuse to distinguish) between life and non-life, sometimes redefining and refashioning *bios* as we know it.

1.1 What, then, is life?

On the microcosmic scale, with the new genetics and associated advances in bioinformatics and engineering, life has become increasingly unstable. Interestingly, as Keller (2003) points out, as biology has become Big Science, escalating the study of life itself on an unprecedented scale, it also has *dissolved* the essence of life, partly through "synthetic biology," the "making" of life from chemical building blocks. With the upgrading of the life sciences to Big Science, then, life has become both a focus of attention and a foggy terrain. Schrödinger's question "what is life?," as a result, has become more tricky and pressing. Or is it beside the point? For many scholars, as Dupré and O'Malley suggest, the border case of viruses highlights the fuzziness of life (see Dupré 2012: 208). While viruses are capable of entering cells and moving DNA from one organism to another, they can neither metabolize in the usual sense nor can they reproduce themselves autonomously. No doubt, the exact definition of life will remain elusive and contested, frustrating taxonomic enthusiasts, even more so than in the past. At the planetary scale, the rhetoric of the "Anthropocene" has rendered humans as a geological force, capable of leaving a massive and lasting imprint on the biosphere of the planet. Perhaps in the future we will speak of "synthetic geology."

The gene talk of the second half of the last century – in the wake of the discovery of the double helix and the mapping of the human genome – reinforced the idea of life as a biological given, as a platform on which social life was established and played out. The successes of the new genetics and the idea of the "secret" or "blueprint" of life both delayed genuine biosocial syntheses and silenced critique of the notion of the biological given. Ingold (2011: 3) suggests that generations of theorists "have been at pains to expunge life from their accounts, or to treat it as merely consequential, the derivative and fragmentary output of patterns, codes, structures or systems, variously defined as genetic or cultural, natural or social." His own work for the past quarter of a century, he suggests, has been driven by an ambition to reverse this emphasis, to "restore anthropology to life," through several avenues, including the phenomenological notion of dwelling and the idea that life is lived along lines.

It is one thing, perhaps, to bring *anthropology* to life, moving beyond codes and scripts, and quite another to bring *life* to anthropology, to think comparatively about life and its diverse manifestations and conceptions (Pitrou 2014). Drawing upon the "biopolitical" works of Foucault, many scholars have usefully directed critical attention to the life sciences, in particular the radical refashioning of living matter and its broad impact on social life. As Fassin (2009: 46) has pointed out, however, in Foucault's biopolitics "'life' remains largely elusive." For "the lived" – "life which is lived through a body (not only through cells) and as a society (not only as species)" – Fassin (2009: 48) proposes the label "life as such," to avoid a narrow focus on "biological" phenomena. Lemke (2014: 8, 11) seeks to qualify Fassin's analysis of Foucault's understanding of life, arguing that in some of his last works on the "government of things" Foucault goes beyond the dualistic and anthropocentric limitations of his social-constructivist works on biopolitics. For the post-human Foucault, "the milieu articulates the link between the natural and the artificial without systematically distinguishing between them"; "the biological can only play out in a certain 'milieu.'" This would resonate, Lemke (2014: 11) suggests, with current forms of vitalism and new materialist scholarship; "life is not a given but depends on conditions of existence within and beyond life processes."

The chapters in this book seek to address different conflated zones of life at particular times and scales, from the genome to the human body and the global environment. Life itself is currently one of the most active zones of politics and economic production, as biological material is increasingly the subject of engineering, banking, reproduction, and exchange. The description and broad implications of these developments represent some of the most challenging issues on the academic agenda at the beginning of the twenty-first century. This is highlighted by numerous recent studies in the social sciences, humanities, and life sciences. Often associated with the "culturing" of life and the biosocial turn, these events suggest new kinds of concepts, theoretical perspectives, and politics, calling for new forms of engagements – and new anthropologies of life.

Speaking of "biosocial relations," I am arguing, helps to capture the hybrid nature of life itself; in particular, it facilitates sensitivity to differences and similarities in hierarchies involving the fashioning of life, the reproduction of bio-objects, exchange between humans and across the species divide, and the co-constitution of humans, other animals, and, more generally, the environment. The discussion moves along broader spheres of life – biomes, ecosystems, and the Earth itself – to illuminate different contexts of life, associated notions of agency, and their place in various forms of environmentalism and biopolitics.

There are good grounds for rethinking the human condition and the Kantian question: "What are human beings?" For one thing, the post-human condition is rapidly advancing, with the growth of artificial intelligence, human prosthetics, and cognitive science. With the new genetics, moreover, what used to be called "life itself" is increasingly modified by humans through artificial means, undermining the separation of the "natural" and the "artificial." As Rabinow (2008: 14) argues, this calls for systematic theoretical reflection and ethnographic documentation: "[T]he *logos* of *bios* is currently in the process of rapid transformation. Therefore, a central question before us today is: given a changing biology, what *logos* is appropriate for *anthropos*?" Often associated with the notion of "biosociality," this turn of events was first documented in the context of feminist studies of reproductive technologies (Martin 1987, Carsten 2004, Franklin 2013). While the new genetics and reproductive technologies firmly placed biosociality on the theoretical agenda, in an important sense biosociality is characteristic for human life (Dupré 2012, Ingold and Palsson 2013). This seems to demand new kinds of concepts for anthropology and related fields.

Speculations about the unity and integrity of the discipline of anthropology have been recycled repeatedly. As early as 1936, Franz Boas remarked that for a few decades physical and social anthropology had been drifting "more and more apart" (Boas 1940: 172). It may seem that, especially with the new genetics, the ruptures within the discipline have become terminal – continental plates, in geological terms – with profound translation problems between subdisciplines, and there are recent cases of departmental fission along these lines in North America, defying the classical four-fields approach. Thus, Segal and Yanagisako (2005) suggest that it is time to question what they call the standard oath of loyalty to the unity of the discipline. Moreover, European biological and social anthropologists continue to operate on different terrains in the disciplinary landscape (Ingold 2001), often speaking past each other. Yet the overall trend does not seem to be toward a complete split along the nature/society divide. Indeed, there are growing demands for reintegrating the natural and the social and for strengthening transdisciplinary forums and collaborations, some of which have genomics-and-society as an important theme.

The trajectories of the two major subdisciplines of social and physical anthropology thus invoke Zeno's paradox of Achilles and the tortoise. Challenging Achilles, a skillful athlete, to a race, the tortoise reasons that it is bound to win as long as Achilles gives it a small head start. Achilles concedes the race, convinced by the argument of the witty tortoise that it would be endless: Once Achilles covered half the distance between them, he would have to cover half the remaining distance, and so on forever. Whereas Achilles and the tortoise imagine that the distance between them will be successively reduced without ever being eliminated, biological and social anthropology seem to refuse to separate despite continually drifting apart for over a century and despite the apparent acceleration of the drift in the wake of genomics. The paradox may be resolved not by logical or mathematical tricks along the lines of the ancient Greek but by the fact that nature and society have been redefined.

One important focus of recent debates on anthropological theory relates to the notion of the biological and the ways in which it – rather arbitrarily, I argue – splits the discipline of anthropology. One of the avenues with important implications for anthropologies of life might be called the "biosocial turn." Given such a turn, anthropology has become an integral part of an expanding network of diciplines that embraces and advances dialogues between social and biological perspectives that have tended to be separated.

1.2 The biosocial turn

One of the early precursors to the concept of the biosocial is Mauss's reference to the "biologico-sociological" in his classic essay "Techniques of the Body" originally published in 1934 (Mauss 1973). For Mauss, the "habitus" represented by acts like walking, swimming, and dancing was both a biological and a sociological phenomenon; movement was normative both because it was bodily inscribed and because it was informed by the traditions of the community involved. Since the 1960s, at least, following the launch of the *Journal of Biosocial Science* in 1969, which replaced the *Eugenics Review* published by the Galton Foundation, the concept of the "biosocial" has often been used loosely with reference to "the common ground between biology and sociology," to quote the journal's home page (Journal of Biosocial Science 2007). In a review of the first issue of the journal, published in *Man*, Roberts (1970: 133) suggested the main problem for the journal would be to establish meaningful common ground, emphasizing that it was "not sufficient merely to put papers on human biology within the same cover as others dealing with the social sciences." A similar concept of the biosocial has been highlighted by The Biosocial Society (2007), an international academic body which "aims to foster closer collaboration between those biological and social sciences engaged in exploring human biological and social diversity."

In the 1970s, the dualism of the social and the biological was sometimes rendered as "biosocial anthropology." For Fox (1975: 2), biosocial anthropology was neo-Darwinian, a branch of evolutionary biology: "[I]t accepts as its premise the role of mutation and natural selection as the main point of departure for the analysis of anything concerning the life processes of any species. It views social behavior, then, as the outcome of an evolutionary process and analyzes it as such; cultural life being an outcome of the same process and only understandable in these terms." Knowledge of what people make of themselves is subsumed under biological knowledge of the human being. "Biosocial analysis, it should be emphasized," Fox (1975: 2) continues, "is an analysis of the interplay between biological 'givens' – whatever their nature – and cultural responses. Many responses are possible, but always to the same givens. What this mode of analysis opens up is a new possibility for getting at the givens and hence a better chance to understand the variation in the responses." Interestingly, this notion of biosociality seems to have largely been forgotten; it hardly leaves a trace in the rapidly growing literature on biosociality in anthropology. Perhaps it testifies to the dual split of the biological and the social wings or subdisciplines.

In these cases, the biosocial (and the "biologico-sociological") refers to two separate relational systems, one biological and the other social, suggesting a dualistic division of academic labor. Inherited from Durkheimian theory, this dualism was underlined in Mauss's work. For him, the notion of the "cogwheel" (Mauss 1973), a reference to some kind of mediating psychological mechanism, ensured the coordination of the two spheres of the biological and the social. While Mauss and several others drew attention to the body, it remained silent or absent-present in social thought; and either it was marginalized or it was subjected to the reductionist gaze of the biological and medical sciences. This is what Ingold (2001: 256) refers to as the "complementary approach," an approach that aims to "put together the partial accounts of human life obtainable to each of the two planes, of nature and society, to produce a complete 'biosocial' picture." The alternative "obviation approach," he suggests, would reject the complementarity assumption "not ... by simply collapsing one side of the dichotomy into the other as in the more extreme forms of socio-biology and social constructivism, but by doing away with the dichotomy itself" (Ingold 2001: 256–257). Franklin (2003: 66) cautions, however, that while it is no longer possible to see the "natural" and the "social" as ontologically different, "the natural facts–social facts distinction may need to be reinvented, rather than discarded, in order to understand the kinds of connections and relations being produced in the context of the new genetics."

A very different notion of "biosociality" from that of Mauss and Fox arrived on the scene in an important essay by Rabinow (1996). In his vision, the conceptual division of nature and culture was about to collapse with the new genetics and the mapping of the human genome, eventually completed soon after the turn of the century. Developing a similar argument, Rheinberger (2000a: 19) suggested that the molecular biology advanced between 1940 and 1970 not only represented a paradigm shift founded on the notion of information, but also gene technology facilitated "the prospects of an intracellular representation of extracellular projects – the potential of 'rewriting' life." Indeed, the root meaning of the word biotechnology is living technology, biological artefacts serving human ends. The traditional dichotomy between "nature" and "culture," then, no longer makes much sense. Arguably, the reality of "biosociality," the conflation of the biological and the social through modern biotechnology, dissolves the earlier concept of the biosocial – the notion of the complementary spheres of biology and society usually seen to underlie the dualistic structure of the discipline of anthropology and, in fact, most academes.

The product of a long process of evolution spanning at least 200,000 years, humans now reinvent themselves in a new sense and on a fundamentally new scale, deliberately altering their bodily constitution and development partly by exchanging genes, tissues, and organs with both conspecifics and other organisms. Foucault's works on biopolitics (see, for instance, Foucault 1994) have obviously contributed critical insights with respect to the current refashioning of the human body, illuminating the political and governmental dimensions of these developments (Rose 2005, Gottweis and Peterson 2008). Recently, a series of scholars have revisited the early writings of Marx, sometimes in combination with Foucauldian perspectives, in their attempt to make sense of the political economy of modern biotechnology, including the fragmenting of body parts and the labor process involved.

Underlining the conflation of the social and the biological, Thacker (2005: 17) has recently argued that with biotechnology human bodily material has been turned into machines: "Using the cut-and-splice techniques of genetic engineering, scientists can insert the human gene into the bacterial plasmid ... As the bacteria replicates, so will the inserted human DNA, making for a kind of biological copy machine." Knorr-Cetina treats laboratory mice used in the production systems of experimental science as "biological machines." To her, the notion of the machine can be used as "a master analogy for the ontology of objects" in the experimental system of the laboratory: "The autonomous production units into which organisms are decomposed ... are *molecular machines*. Other materials in the lab may not function on a molecular level, but they are still used and usable as *biological machines*" (Knorr-Cetina 1999: 149; emphasis in the original).

Like many other anthropologists, Lévi-Strauss was concerned by the neo-Darwinian reduction of social practices and institutions to evolutionary processes of selection, fitness, and the like. Given the enormous theoretical significance he attributed to the nature/culture divide, the key binary opposition in his structuralism, one might not, perhaps, expect Lévi-Strauss to be prepared to go beyond it. Nevertheless, he seems to have sensed the destabilization of the nature/culture divide itself in the wake of the new genetics. When pressed about the implications of genetic discoveries and the extent to which they might eliminate the distinction between nature and culture, he responded that the distinction maintains its value in that it "provides a barrier against those offensives, such as sociobiology, made by simplistic and limited minds, that would have cultural phenomena reduced to models copied from zoology" (Lévi-Strauss and Eribon 1988: 106). However, he adds an important qualification:

> If one day the boundary between nature and culture vanishes, it won't be along what we refer to today as the interface between human and animal phenomena, i.e., there where certain human characteristics, such as aggression, seem to resemble what is observed in the behavior of other species. *If this change takes place, it will occur elsewhere, involving the most elementary and fundamental mechanisms of life and the most complex human phenomena*. If the boundary is to disappear it will be behind the scenes where partisans of culture and nature are presently debating. (Lévi-Strauss and Eribon 1988: 106; emphasis added)

For a long time, anthropologists have pointed out, drawing upon their ethnographies from non-Western contexts, that the nature/culture opposition is not a universal one. Although the Hageners of Papua New Guinea, Strathern (1980) argued, *did* make a distinction between the wild and the domestic, that distinction did not seem to carry the main connotations usually applied to nature/culture discourse, including the idea of natural law and human mastery. More recently, some anthropologists have argued that while dualism may be evident in some non-Western contexts, it may take radically different forms. Thus, Viveiros de Castro (1998) suggests the term "multinaturalism" to capture the essence of Amerindian conceptions, in contrast to the multi*culturalism* of Western cosmologies. Amerindian concepts, he suggests, reverse the key axis of modernist thought by setting human culture, not nature, as the universal or the *a priori*, assuming that *nature* is differentially constructed by cultural subjects. Perhaps, the Amerindian perspective of multinaturalism testifies to the resonance of many "indigenous" views with the recent notion of biosociality.

The nature/society divide in social and biological theory has been heavily theorized and criticized for decades. For some scholars, as a result, "biosocial" thinking is simply old wine in new bottles. It would be a mistake, however, to dismiss current thinking along these lines as mere repetition. A broad biosocial momentum seems to be taking place across academe, in both the social and the life sciences. The deep-rooted antagonism of the social and the life sciences, each of which has sought to debunk or colonize the other, has somewhat surprisingly and rather quietly given way to an open collaborative zone that renders the nature/society divide utterly trivial and out of place. As Meloni (2014: 3) puts it, "the two extreme wings of the nature/nurture dichotomy are equally destabilized by the new biosocial terrain." Drawing upon recent developments in neuroscience, genomics, epigenetics, and social studies of race and inequality, he concludes that "the contemporary presence of genuine conceptual transformations in so many disciplines is unprecedented and has never been favoured by scientific evidence to the extent it is today" (Meloni 2014: 11). For Meloni, the "extreme sociality of biology" is not the product of some cultural logic but "*a realist sociality*, so to speak, something that is intrinsically part of the functioning of the facts of life" (2014: 13; emphasis in the original).

One sign of the current biosocial momentum is the growing dissatisfaction with the interactionist rhetoric of gene–environment, body–society, and nature–nurture. The language of interactions, it is argued, misconstrues life itself, artificially separating in advance the domains that supposedly interact in the process of life. In Ingold's words:

> That life unfolds as a tapestry of mutually conditioning relations may be summed up in a single word, *social*. All life, in this sense, is social. Yet all life, too, is *biological*, in the sense that it entails processes of organic growth and decomposition, metabolism and respiration ... It follows that every trajectory of becoming issues forth within a field that is intrinsically social and biological, or in short, *biosocial* ... This is why we speak of humans ... not as species beings but as biosocial becomings ... *The domains of the social and the biological are one and the same.* (Ingold 2013: 9; emphasis in the original)

A similar idea that moves beyond the interactionist perspective is that of "body worlds" in some recent archaeological works. The common-sense view of the body as a natural, physical object eventually dressed up in "culture," Harris and Robb (2013b) argue, is not only a historical product, emerging with the "body as machine" in the sixteenth century, it is also seriously flawed and incomplete in that a universal "natural" body doesn't pre-exist the "social" body: "[H]umans can never not be social. The very structures and processes of the physical body itself always develop within social relations" (Harris and Robb 2013b: 213). The fact that body worlds differ with time and place doesn't mean that there is "a single 'real' (biological) body" (Harris and Robb 2013b: 215); the body is biosocial throughout.

In some ways, the framework of biosociality captures what is often referred to as "nature-cultures" in the literature. Some interpretations of the latter, however, suggest a critical difference that needs to be addressed. Goodman (2013: 361) rightly points out that "the division between biology, the humanities, and the social sciences has reinforced an understanding of bodies as unchanging natural entities." The notion, however, of "cultural-biologicals" which he advances to capture "the cultural that is always in human biology" seems to assume the existence of a "pure" biological substrate, some kind of bare life, that becomes infused with culture through practice: Such a notion, Goodman argues, will advance "appreciation of how the local *gets into bodies* and becomes biological" (2013: 360; emphasis added). The "local" is always necessarily there.

1.3 Biosocial relations

As I have argued elsewhere (Palsson 2013), humans can be seen as ensembles of biosocial relations. Human becoming, in fact, is a thoroughly relational, biosocial phenomenon – collective history embodied and endlessly refashioned in the *habitus*. Race, gender, social class, and other themes on human variation are social as well as biological, embodied signatures of human relations and histories, indicators of degrees of well-being (Bliss 2012). "Biology," in such a broad biosocial sense, is destiny. This is not, however, to succumb to dual determinisms, one social and another biological. Humans fashion their lives and contexts through their agency, practices, and politics. Fortenberry (2013: 165) suggests, in the context of the significance of microbiomes, that rather than avoiding the issue of "color" (becoming colorblind) microbiome researchers should closely attend to it since it is likely to reflect racial disparities in health: "[T]he microbiome becomes a critical tool for understanding the pervasive influences of inequality that are the social, psychophysiologic, and environmental contexts that link racial/ethnic categories and health."

An extended notion of social relations of production may be useful for capturing new hierarchies and articulations of the social and the biological in the reproduction of life itself, what might be called *biosocial relations of production*. Coupled with detailed ethnographies of biomedicine and the bioindustry, such an extension may serve to highlight the micropolitics of what Marx referred to as living labor. While Marxian rhetoric has often been at odds with ethnographic description, it seems to make good sense to apply Marx's notion of mode of production to the fragmenting and co-constitution of bodies and the reproduction of bodily material. A Marxian approach along these lines is already in the air. Several important works have drawn upon Marx's concepts of labor, estrangement, and species-being; see, for instance, Thacker (2005), Thompson (2005), Sunder Rajan (2006), Dickenson (2007), and Haraway (2008). Even Derrida (1994: 67), the arch-deconstructivist, conceded in his *Specters of Marx*: "The critical treatment to which ... [Marx] subjects the abstract concepts of Nature and Man as man remains a rich and fertile one." Some of Marx's notions are surprisingly relevant, almost hypermodern. Applying them to the novel domain of biotechnology, however, needs some qualifications and fine tuning.

One of the hybrid developments that sparked my writing on these lines was the birth of seven "sensational" pigs at the Foulum Research Centre in Denmark, reported in the daily *Politiken* in August 2007; apparently these were the first "Alzheimer's pigs" ever, the result of cloning and genetic manipulation, with an added human gene implicated in the onset of Alzheimer's disease. The seven pigs (a magical number, indeed), it was hoped, would develop symptoms similar to those experienced by Alzheimer's patients, providing new opportunities for researchers to explore brain tissue at different stages of development. Perhaps, the comparison of human and animal slavery, as a result, needs some rethinking. At least, the novelty of such cases would fly in the face of much classical social theory – for instance, Durkheim's thesis of totemic associations of animals and humans. For him, the totemism of Australian "primitives" represented dubious analogies between people and certain animals: "[T]here is nothing in experience which could suggest these connections and confusions. As far as the observation of the senses is able to go, everything is different and disconnected. Nowhere do we see things mixing their natures and metamorphosing themselves into each other" (Durkheim 1971: 235–236). We do, indeed, "see things mixing," as the case of the Alzheimer's pigs demonstrates. In Marxian terms, this is lively production, the collaborative project of pigs and humans.

There is good reason, I suggest, drawing upon Haraway (2008) and some others, to extend the notions of production and estrangement to the realm of human–animal relations, in particular the role of non-human animals in experiments involving human diseases and the development of "spare parts" for human use. As Haraway (2008: 46) puts it, "What ... if *human* labor power turns out to be only part of the story of lived capital? ... [W]hat if the commodities of interest to those who live within the regime of Lively Capital cannot be understood within the categories of the natural and the social that Marx came so close to reworking but was finally unable to do under the goad of human exceptionalism?" In fact, the Marxian approach to human production systems has sometimes been applied in an extended sense to human–animal relations of production. Such an approach was developed by Tapper (1988) to illuminate both the different ways in which animality and humanity are socially constructed and the different hierarchies involved in human–animal production systems:

> A Marxian classification of social and economic systems by *mode of production* is not apposite, since its central component, comprising *human* social relations of production, does not take account of relations of production between humans and animals . . . More useful . . . is to cast a Marxian frame around the classic typology of production systems, which *are* characterized by specific human–animal relations of production. These systems are hunting and gathering, pastoralism, agriculture and urban-industrial production. (Tapper 1988: 52; emphasis in the original)

Tapper's comparison of these systems underlines that hunter-gatherers, unlike most other producers, typically live in complementary relations with the other animal species in their environment, often describing exchanges with them in terms of an ethos of reciprocal, co-operative exchanges. Sometimes, hunters tame particular animals (such as reindeer), taking individual animals out of their natural species community to provide labor for humans and treating them as slaves. In such "ancient" systems of production, which also characterize cultivators who use draught animals, the reproduction of the animals is under the control of their human masters. Among pastoralists, in contrast, production is based on animals that are not tamed but are herded in communities; while the herds are monitored and managed by their human masters, the relationship is "like a contract or transaction in which the masters 'protect' the herds in return for a 'rent.'" This resembles the Marxian conception of feudal relations between lord and serfs" (Tapper 1988: 53). In the modern form of pastoralism, in ranching, to provide one more contrast, animals are herded in large numbers with no personal relations with the owner of the ranch: "These seem . . . to be typical – paradoxically for a modern offshoot of capitalism – of Asiatic-Oriental relations of production. Indeed, the cattle 'barons' of the Texas ranges should perhaps be termed 'sultans' – or 'moguls', like their oil-rich successors" (Tapper 1988: 53). Finally, in urban-industrial society, in battery- or factory-farming, animals are reduced to machines and exploited along classic capitalist lines (Lien 2015). Such an approach underlines the hierarchies of human–animal production systems, using a variety of terms – reciprocity, co-operation, slavery, contract, protection, and exploitation – that focus on one or more aspects of the system: the human producer, the animal, and the relation involved.

An extension of Tapper's original thesis beyond its "natural" production domain to that of cross-species projects in biomedicine and biotechnology seems pertinent. Not only are xenotransplantations routine operations, with humans on the receiving end of organ transplants, human genes are nowadays increasingly introduced into alien bodies – in particular, pigs – for the purpose of studying the onset and development of "human" diseases. One example is the "Alzheimer's pigs" previously mentioned. While laboratory animals are subservient to humans in both xenotransplantation and gene transfer, these two contexts position pigs in radically different kinds of biosocial relations to humans. In the former they are raised to produce "spare parts" to be inserted into human bodies for repair, while in the latter their bodies operate as both surrogates for human body parts and living laboratories for exploring malfunctioning human bodies. Presumably, the former are manufactured on a rather large scale, by some kind of "sultan" or "baron," while the latter are produced and raised in small numbers with greater attention to individuality, care, and detail. At any rate, while the animals employed by biomedical laboratories may produce "one-sidedly," for a specific *human* purpose, they hardly do so in the Marxian sense of producing only themselves.

The notion of relations of production, it may be noted, has been applied with somewhat similar aims in radically different contexts – for instance, that of textual studies and translation theory. Thus, Lefevere and Bassnett (1990: 11) emphasize the relation of power between source (original text) and receptor (translation): "[A]lthough idealistically translation may be perceived as a perfect marriage between two different (con)texts, ... in practice translation takes place on a vertical axis rather than a horizontal one. In other words, either the translator regards the task at hand as rising to the level of the source text and its author or ... the translator regards the target culture as greater and effectively colonizes the source text." Sensitivity to relations and hierarchies, I suggest, is essential for "thick" descriptions of new forms of life and, by extension, for informed biopolitics and governance. What, then, should the reference to biosocial relations of production be taken to mean and how might it be applied to bodies and their disembedded products?

At one point Marx speculated on the production zone most likely to generate dynamic economic developments, comparing the tropics and temperate regions: "A too prodigal nature," he suggested, "fails to make man's own development a nature-imposed necessity. It is not the tropics with their luxuriant vegetation, but the temperate zone, that is the mother-country of capital" (1976: 513). The "mother-country" of modern biocapital, of course, is nothing less than life itself. Such a notion, in fact, is more pertinent than Marx could possibly have anticipated, given the central importance of the human body in modern bioindustries, the relative role of life, in contrast to the dead labor of machines. Thus, Thacker (2005: 182) suggests that "Marx's distinction between living labour and dead labour be taken quite literally. Living labour in the biotech industry is, quite simply, 'life itself.'"

1.4 Moving concepts

Rabinow did not elaborate on the meaning and potential usefulness of the biosociality concept. Whatever his original intentions and motivations, the concept took on a life of its own. While nowadays it has become an established part of the vocabulary of students of the humanities and the social sciences focusing on the new genetics, testifying to some kind of usefulness, different authors are not necessarily operating with identical ideas. For some, biosociality refers to changing notions of identity and belonging in the wake of the new genetics, in particular the ways in which people organize themselves into groups on the basis of emerging evidence of the genetic risk of developing a disease, tracking down relatives and people with a similar predicament, and lobbying for research and, possibly the development of drugs or other remedies. In Thompson's words, "[t]he notion of peoplehood that scholars of medical technologies have coined and begun using is 'biosociality'" (2005: 252). Such forms of biosociality depend on a host of factors – among other things, access to the Internet and local notions of health, medicine, personhood, and expertise. As Rose (2005: 147) points out, the kinds of biosociality found in the United States, Europe, and Australia reflect particular conceptions of citizenship and personhood: "[S]uch forms . . . have no visible presence in many geographical regions. AIDS biosociality in sub-Saharan Africa is very different from that of Paris, San Francisco, or London."

Hacking (2006: 81), no doubt, is right in suggesting that the notion of biosocial identities has appealed to many students of biotechnology in recent years: "Currently, the genetic imperative – the drive to find biological, but above all genetic, underpinnings for all things human, in sickness or in health, in success or in strife – is fueling fascination with this concept." On the other hand, for a growing number of scholars the usefulness of "biosociality" seems much broader, extending far beyond identity and belonging. In their examination of the literature, Gibbon and Novas (2007) both identify the key conceptual arenas where the biosociality concept has gained currency and explore how the concept may be put to work in new ways. While biosociality, they suggest, has often been used in reference to the identity practices already mentioned, it has also been applied both to the reframing of the nature/culture divide and in the context of emergent and unfolding arenas of scientific inquiry. It may be difficult to avoid the dualistic traps of the early language of the biosocial; thus, the twin notion of "biologies" and "socialities" seems to be a tempting alternative. However, the refiguring of life itself, the reality of biosociality, necessarily destabilizes such dualism.

Rheinberger (2000a: 29) argues that with molecular biology and gene technology we have become "aware that we live in a world of hybrids for the characterization of which we run short of categories." I suspect that to many people the category of biosociality, along with several others, has served exactly that purpose, of capturing some of the hybrids of modern biotechnology. Indeed, reflecting on his concept fifteen years after its launch, Rabinow suggests that "the question was: how had sociality changed given the rise of the new understandings of genetics? Thus, the term biosociality was coined as an initial attempt at *framing the issue of re-problematization of 'life'*" (2007: 188; emphasis added). It seems pertinent to speak of biosocial relations of production to capture the biosocialities involved, the different materialities and hierarchies of the political economy of the fragmented body. The point, of course, is not to construct a tidy and rigid classificatory scheme, but to facilitate sensitivity to differences and similarities.

While my discussion has sought to outline in fairly general terms an important territory for further theoretical exploration and empirical description, drawing upon the insights of several scholars representing several disciplines, at this stage some qualifications are needed. For one thing, the Marxian notion of labor may be extended too far in some of the recent works experimenting with Marxissant vocabulary. Thacker (2005: 300) argues that "Marx's conditions of labor power have been rewritten by the biotech industry"; now, he suggests, it is not "the human worker, who views his or her labor power as property to sell, exchange, and circulate" but "a nonhuman biological network of cell lines, tissue cultures, and genomic databases. Labor is not, then, real-time labor of the physical body; instead it is the archival labor of cell cultures, databases, and plasmic libraries." Clearly, cell cultures, databases, and plasmic libraries do valuable work, but to see such work as "labor" seems to presuppose consciousness of a relationship to that which is being produced, given Marxian theory (see, for instance, Arendt 1958: 96–101), which is hardly the case for biosocial assemblies of this kind. Perhaps it makes sense in some contexts to speak of "production" rather than "labor."

On a related score, the issue of alienation and estrangement needs to be theorized and explored more closely in the context of biotechnology, given the different mediations and circulations of bodily material in the biosocial process of production. While the extension of such notions, I suggest, to the extraction of some bodily material (organs, for instance) is obviously illuminating, with increased distance from the production site (in the case of tissues, cell lines, and databases, for instance) claims about alienation and estrangement become progressively less persuasive. Arguably, moreover, the biosocial relations of production sometimes *render* a "thing" as "mine" that wasn't property in the first place, resulting in a subjective feeling of alienation and estrangement. This seems to hold, for example, for the collection of cell lines from "indigenous" communities, sometimes resulting in charges of biopiracy. Some scholars have theorized, one may add, that with the New Economy "we may be witnessing the end of property in the person, that is, the end of modern notions of personhood" (Adkins 2005: 126). Such an argument seems to have important implications for the understanding of a number of aspects pertaining to the kinds of biosocial relations discussed here, including those of the concepts of labor, gender identities, and alienation and estrangement. If it no longer makes sense to speak of property in the person, does it make sense to speak of the alienation and estrangement of body parts?

A further qualification relates to the temporality of biotechnology. Focusing on the industrializing of organisms, Russell (2004: 9) suggests that their capacity to work, much like that of the human laborer, is always limited by particular lively properties and particular biographies. While this is an interesting and potentially useful perspective, it obscures an important aspect of certain kinds of biotechnologies. As Landecker (2007: 11) emphasizes, the development of laboratory tissue culture implies *regulating* cellular time, manipulating biographies: "[C]ells freed from the bounds of the body are also freed from the limits of the originating organism's lifespan."

Decades ago, Bennett (1976: 4) suggested that the concept of "human ecology" was a myth; due to the "growing absorption of the physical environment into the cognitively defined world of human events and actions," he reasoned, "there is (or shortly will be) only, and simply, Human Society: people and their wants, and the means of satisfying them." The recent development of biotechnology and the industrializing of organisms make such a statement even more pertinent than before. Not only is the "physical environment" increasingly rendered as human construction; life itself is a biosocial artefact. The mother country of biocapital, as we have seen, is characterized by the manufacture of many kinds of "natures" involving a variety of bodily exchanges among humans and between humans and other animals. Several social theorists have argued for a constitutive model of the "person," underlining that individuality necessarily presupposes involvement in social relations. For Marx, for instance, the individual is "an ensemble of social relations" (Marx and Engels 1970: 122). Likewise, drawing upon Melanesian ethnography, Strathern (1996) has theorized the notion of the "dividual" person, an aggregate of networks and relations. It makes sense, I think, to talk about the person as a dividual ensemble of *biosocial* relations.

It would be too idealistic, however, to say that humans, at last, are mastering nature; this would mean lapsing into the modernist framework that seems to have more or less crumbled under the pressures of advancing evidence of biosociality. Surely, however, life itself is being intentionally refashioned, possibly relegating evolution to the back seat. This turn of events not only suggests revised division of academic labor, post-disciplinary collaboration across the now suspect nature/society divide; it also demands new kinds of concepts, politics, and ethics. For a growing number of scholars, the notion of biosociality captures these developments, undermining early dualistic notions of the biosocial. The task remains to systematically chart the bewildering complexity of relations, hybrids, and hierarchies in the making, to explore how the agents involved understand them, and to unpack what they might mean in the broadest sense for both contemporary and future biosocial life. Some of this complexity is explored in the following chapters.

I

Selves

The chapters in this section focus on gene talk and name talk and their uses as devices for exclusion and belonging and as explanations of social life. Also, the section addresses large-scale personal genomics projects and services, emphasizing the role of users. The discussion of belonging and inheritance is partly framed within the context of research in Nunavut (Canada) and Greenland, emphasizing the similarities and differences in scientific discussions of the constitution of the individual through the vertical transfer of substance and Inuit notions of personhood, subjectivity, and relatedness. Genomic anthropology has played an important role in this domain, drawing upon and expanding anthropological understanding of human genomes and their differences. Focusing on deCODEme, 23andMe, and similar projects, which offer personal services to people who wish to assess genetic risks for common diseases and to explore geographies of ancestry, I discuss the difficulties encountered by such projects and their larger social implications. I argue that the users of genomic "stuff" not only perform labor on themselves, seeking to know themselves and to care for their bodies, their labor also needs to be situated in the mode of production involved.

2 Spitting image: decode me!

> **spitting image, spit 'n' image.** *Informal.* exact likeness; . . . bef. 950; (v.) ME *spitten,* OE *spittan;* c. G (dial.) *spitzen* to spit; akin to OE *spætan* to spit, *spætl* spittle . . .
>
> Dictionary.com, 2009[1]

In 2008, the California-based project 23andMe – a project that offers to estimate a person's predisposition to a number of traits and diseases on the basis of a saliva test – held a "spit party" during New York fashion week; volunteers would spit into a test tube to provide their DNA for sequencing and analysis. This was reported as a vibrant scene, a young couple opening their kits and donating saliva, to explore what their genomic constitution might tell them about their identity and the kind of life they might lead. Apparently, they were publicly celebrating both their self and their genome, staging their persona and their bodily essence for the media, in the process of lobbying for personal genomics and the company responsible for 23andMe, an affiliate of Google. An article in the *New York Times* announced the launch event by saying that 23andMe "wants people to think of their genomes as a basis for social networking," adding that "the company . . . hopes to make spitting into a test tube as stylish as ordering a ginger martini" (Salkin 2008). In November 2008, *Time Magazine* declared the retail DNA test of 23andMe the best innovation of the year. The year before, Apple's iPhone was the winner. Several other companies have started one form or another of retail genomics. This is consuming genomics, a rapidly growing business receiving both substantial financial support and intense public attention. Clearly, something new was in the air.

The notion of spitting and related concepts has proved to be a powerful metaphor. Exploring its social history is like following the trajectories of ancient DNA. The English verb to "spit" – to "spew" or to "expel saliva" – is of early medieval origin. The noun "spit," in the sense of "the very likeness," is more recent, attested from 1602, while "spitting image"

[1] Dictionary.com, 2009, http://dictionary.reference.com/browse/spitting. Accessed May 29.

is a twentieth-century thing, apparently from as early as 1901. It may be interesting to note that there has been some debate on the etymology of the phrase. Some have suggested it is derived from "*splitting* image," based on the two identical parts of a split plank of wood. Such an account would resonate perfectly with the modern concept of the double helix and the splitting of DNA, underlining the relevance of the idea of the "spitting image" for both modern gene talk and the genealogical tree. The discovery of the structure of DNA material has been heralded as the key to the understanding of the continuity and change of organisms and life forms, as the missing conceptual link of evolutionary theory finally solving the mystery of the "tree of life." Also, "splitting" might highlight the Western notion of the duality of the individual as a natural body and a social person, a notion often challenged nowadays by the monistic concept of the biosocial. Given such reasoning, the 23andMe spit party might just as well have been called a "split party." It seems, however, that the reference to spitting was based on "spit," not "split," an allusion to someone who is so similar to another as to appear to have been spat out of his or her mouth (Martin 2009).

The spit party and the notion of spitting image invite interesting anthropological questions: What are the overall spin-offs from personal genomics, especially with respect to the understanding of self, personhood, relationships, and ancestry? Despite sustained criticisms of the gene talk current in the West and the determinisms it implies, personal genomics along the lines of 23andMe seems to have a substantial public appeal. At the same time, the services offered by genomics companies give rise to new kinds of relations and networks based on genetic signatures presumed to be encoded in DNA. Like many others, I decided to indulge in a kind of spitting, mixing ethnographic observation, theoretical reflection, and narcissistic pleasure. One of the key companies in the development of personal genomics, deCODE genetics, happened to be located on the outskirts of my campus in Reykjavik, Iceland – within spitting distance, if you like.

Anthropology has played an important role in both personal genomics and studies of human variation, drawing on and expanding anthropological understanding of human genomes and their differences. I shall argue that in the process anthropology has coproduced new biosocial networks of associations along with a whole series of technologies and agencies engaged with biomedical research (Lock and Nguyen 2010). At the same time, the boundary between experts and laypersons has been blurred and refashioned. Projects such as 23andMe, deCODEme, Pathway Genomics, and Navigenics offer test kits for a low price (ranging from $250 to $2,500) – within the reach of the public, not just the research elite and the wealthy. Analyses and interpretations of genome scans are now a matter of intense public discussion through all kinds of media, including Web browsers and blog sites.

I would argue that up to a point, personal genomics has democratized genomic discourse. However, I also suggest that it is essential to attend to the biosocial relations of production involved, the potential hierarchies in the making in the assembly of personal genomics material and information through which consumers become active collaborators. The new genetics has not only called for the notion of the biosocial, advances in biomedicine and bioscience also make it pertinent to address the broad context of biosocial relations and labor processes (Dickenson 2007).

2.1 From physical anthropology to molecular anthropology

One useful avenue into early twentieth-century physical anthropology is the work of Ernest Albert Hooton at Harvard University. While his views were somewhat difficult to specify – he criticized Nazi physical anthropology, but at the same time he sponsored racist projects (see Marks 2008: 244) – he was one of the most prominent physical anthropologists in America. His work *Up from the Ape*, originally published in 1931, emphasized the continuity of human reasoning on anatomical difference.

> Actually, science is forced to recognize the differences in physical characteristics between the great divisions of mankind... For, the ordinary layman – the plain, untutored Homo sapiens – today and for the past thousands of years has observed these gross anatomical differences between the principal groups of this kind, has drawn the generally correct inference that they are transmitted from parents to children, and has attributed to them enormous political, sociological, psychological, and biological significance, rightly or wrongly. (Hooton 1946: 440–441)

While Hooton emphasized the antiquity of the theme of variability, he was eager to establish the autonomy of his scientific discipline with respect to the ignorant public. Commenting on race, he suggested that "confusions of usage are usually confined to the non-anthropological writing public. All anthropologists agree that the criteria of race are physical characteristics" (Hooton 1946: 447). At the same time, Hooton was keen to set his kind of science apart, irritated by challenges from social scientists.

> Man is an inveterate amateur of the taxonomy of his own kind. He cannot be argued out of the habit of connecting the physical differences he sees in individuals or groups with their equally obvious variations in behavior by any set of "social scientists," however loudly and persistently they tell him that there is no difference between black and white skins apart from exposure to the sun, and no difference between the psychology of a Mongolian and a White, apart from their having grown up in the Rice Bowl and the Dust Bowl, respectively. (Hooton 1946: 447)

For Hooton, skeletal material was the main source of information on human variability. The measurement and classification of bones became an obsession.

This is underlined by the extensive cross-cultural bone collections of the Peabody Museum at Harvard University, a kind of ethnographic atlas engraved in human bones. One of the students at Harvard, the Canadian-Icelandic anthropologist-explorer Vilhjalmur Stefansson, who contributed to the Peabody collection with a sample of medieval Icelanders, later did extensive ethnographic fieldwork among Canadian Inuit (between 1907 and 1918). A frequent phrase in his notebooks is "measured a few heads today" (Palsson 2005: 134, 204). Stefansson would move from one Inuit camp to another, lining up his "noble savages" to have them photographed and measured, setting up what he might have called measurement parties, precursors to the spit parties of 23andMe.

While the measurements of the bone people were exceedingly detailed, their classificatory scheme was highly simplistic and archaic, often with an implicit racial tone. Inevitably, the arrival of human genetics and biological anthropology after World War II provided a new avenue into the understanding of human variability, moving the kinds of measurements and classifications of skeletal material typically practiced by physical anthropologists during the first half of the twentieth century to the sidelines. Did they represent new paradigmatic "thought styles," in Fleck's sense? How radical were the breaks represented by the discovery of the double helix, the mapping of the human genome, the IT revolution and the Internet, the new economy, and related events and developments? Are we able to meaningfully situate ourselves and to domesticate our biases? Interestingly, given the context of skeletal analyses, one of the key illustrations in Fleck's book, originally published in 1935, demonstrates the usefulness of his thought-style perspective through discussion of changing understandings of human anatomy. "To obtain an even clearer picture of how scientific observation differs when two different thought styles are involved," he suggests, "it is perhaps appropriate to compare anatomical descriptions and illustrations in early and recent text books" (Fleck 1979: 133).

As Glick (2008: 240) remarks, the "great debate over race passed the Darwinian divide with scarcely an acknowledgment that anything had changed. Race was an issue marked by lack of ontological control, and that control would be gained slowly and with continuing conceptual difficulty until 'population thinking' introduced some clarity after World War II." The anthropological terrain of human variability in the wake of the new genetics, however, proved to be highly differentiated and rapidly changing, with both radical innovations in methods and perspectives and surprising continuities. Sommer (2008) emphasizes successive contests, following the birth of what Zuckerkandl identified as "molecular anthropology," over what counts as a legitimate epistemic object and authoritative information in the reconstruction of hominid evolution and human variation. The early molecular anthropologists became convinced of the intrinsic superiority and mathematical precision of direct molecular data in comparison with the subjectivity of readings of anatomic data and the fossil record. For instance, at Berkeley, Washburn "heralded the new technologies as finally providing some scientific base to claims about human evolution," ridiculing "comparative anatomy as a kind of pseudo-science that had given rise to just-so stories" (Sommer 2008: 502). Over time, it was assumed, molecular studies would reduce if not eliminate the endemic bias involved when a species was studying itself informed by the belief in human exception. Ironically, however, in due course the analysts became overwhelmed by their faith in their methods. Vincent Sarich, for instance, one of those Washburn recruited at Berkeley, eventually argued against the mantra of "direct evidence" on the basis of molecular data, suggesting that "one no longer has the option of considering a fossil specimen older than about eight million years a hominid *no matter how it looks like*" (cited in Sommer 2008: 504; emphasis added). The thought style of molecular studies had become hegemonic, avoiding contradictory evidence.

2.2 Personal genomics, via Oxford and Reykjavik

The archaeology of the kind of personal genomics we now have on the horizon has several layers, among them maps of human genome diversity, population biobanks, digital genealogies, and the Internet. One of the pioneers of the genetics of ancestry and its commercialization through the Internet is Bryan Sykes of Oxford University, a human geneticist who founded the genetics testing firm Oxford Ancestors, probably the first service of its kind. His company offers people an opportunity to see which "clan" they belong to, tracing their ancestry to one of the seven daughters of Eve (Sykes 2001). To attract customers, Sykes (2001: 197) dramatizes the saga of the seven daughters: "What were they like, these women to whom almost everyone in Europe is connected by an unbroken, almost umbilical thread reaching back into the deep past?" For him, the power of DNA consists in the "token or a symbol of the shared ancestry it reveals rather than the body chemistry it directly controls" (Sykes 2001: 290). Common membership in a clan establishes a profound connection: "We look at each other and sense our deep umbilical connection . . . I feel we have *something very deep* in common" (Sykes 2001: 289; emphasis added).

One of the interesting sites of biomedical experimenting and personal genomics over the past decade is Iceland. Here as elsewhere, the introduction of human genetics represented a change in thought style, with new players and perspectives. Soon after World War II, physical anthropology developed at the University of Iceland through the work of Jón Steffensen, professor of medicine, and Jens Ó.P. Pálsson (no relation to me), founder of the Institute of

Anthropology (see, for example, Pálsson 1976). Both of them worked on skeletal material. Their interests, however, differed significantly (Palsson and Guðbjörnsson 2011); Steffensen emphasized medical and cultural issues without any hint at racial issues, whereas Pálsson tended to draw on public discourses on racial differences and "Nordic" people.

Jens Ó.P. Pálsson partly studied in Germany, where he became affiliated with the Mainz Institute of Anthropology. In Mainz he not only found a place for his fascination with racial types but also received financial support. In particular, he was supported by Ilse Schwidetzky, head of the institute, who was keen to get access to Icelandic data. Despite skillful lobbying in Iceland, decades of data collection, teams of collaborators, and considerable national and international funding, Pálsson's legacy remains small. Thus, he never taught or mentored anybody to speak of. Also, his publication record was meager. Moreover, his text was devoid of theory, focusing on classification and catalogs in the fashion of German physical anthropology, particularly the Breslauer or Breslauer/Mainzer Schule (Preuß 2009: 129). Perhaps he felt at the end of his career, at a time of rapid advances in human genetics, that his work was rather suddenly out of touch, a thought style that failed to sustain attention. Perhaps, too, he had also come to realize that the close connections to Schwidetzky and her entourage at Mainz, a branch of anthropology that remained publicly implicated with the Nazi past, isolated him both internationally and at home. Schwidetzky's anthropology was firmly grounded in the measurements of the Breslauer/Mainzer Schule and the racial typology of Egon Freiherr von Eickstedts, the leading racial theorist of Nazi Germany (see Preuß 2009: 132–134).

The deCODE projects draw on the development of flexible interdisciplinary research teams, powerful genomic laboratories, bioinformatics frameworks, and digital genealogies (Palsson 2007), all of which were involved in the making of deCODEme. The project was able to draw on several kinds of assets – in particular, large-scale efforts over several years to discover the genetic factors involved in common diseases and extensive genomic anthropological work on human populations and their migrations, ancestry, and mixing. All of this was important for developing the analyses and interactive frameworks offered by deCODEme. For a few years, the project offered both a "complete" scan ($2,000) and two narrower scans focusing on specific conditions – cancer ($500) and cardiovascular problems ($500) – but its services have now been discontinued.

2.3 deCODEme: a somewhat personal guided tour

I signed up for the complete scan, curious to find out how anthropological expertise was implicated in the project, to explore the analyses it offers, and to see what the scan might tell me about myself and my roots. Two weeks after I sent my cheek swabs and the relevant forms, I received an e-mail from the company. The results were now available, and I would be able to access them through the password provided. For a few years, I regularly received messages from the company alerting me to both updated and new conditions and to further analyses of traits and health risks. Once I logged on to see the results, I was urged to "have fun browsing [my] ... genome," "dig into [my] ... DNA," explore my ancestry and my "genetic risks," play with maps and other visuals, search for specific genetic variants (SNPs or "snips"), and download my genotypes for 1.2 million SNPs (a 33-Mb data file).

The search for ancestry had six key features. The first, the "atlas," provided a comparison of one's genetic code with that of people from all over the world based on several hundred thousand genetic variants and more than a thousand reference individuals from fifty different populations worldwide. The atlas compared my genome with reference populations throughout the world, ranking regional clusters (1 to 6) in terms of their relevance to me, in order of genetic similarity: Europe (1), Southwest Asia (2), East Asia (3), America (4), Oceania (5), and Africa (6). In each case, I could zoom in on the population involved.

My genome, not surprisingly, turned out to have most in common with European reference groups (a genetic similarity of 83.99 percent), in particular those of Iceland, the Orkney Islands, France, and Russia. More astonishingly, the second feature, "ancestral origins," indicated that judging from chromosomes 1–22, my ancestry is no less than 7 percent East Asian, 16 percent according to the X chromosome, considerably higher than for most Icelanders. I found this an interesting and puzzling revelation. To speak of "genealogical dis-ease" (Rapp, Heath, and Taussig 2001) – to use a term developed by anthropologists studying what people make of genetic information about their roots and ancestry – would, however, be an overstatement.

The analysis of mtDNA establishes one's place in a matrilineal family tree spanning 170,000 years. It turned out I belong to "mitogroup R*," a category shared by 4.8 percent of deCODEme users, all of whom can trace their mtDNA to a woman thought to have lived about 60,000 years ago, probably somewhere in the Near East. The analysis of my paternal DNA, on the other hand, showed that I belong to "Y-group R1a," a category shared with 10.3 percent of deCODEme users tracing their Y chromosome back to one man who is thought to have lived about 10,000 to 15,000 years ago, probably in Western Asia.

A further feature allowed users to explore their "map of kinship," a visual representation of genetic space. Given this evidence, I occupy a somewhat marginal position, neither firmly within the European reference group nor any of the others, probably reflecting the puzzling observation mentioned above about my East Asian ancestry. The final feature allowed the user to compare his or her genome with that of a reference individual from any of the populations included in the data set.

The other main service offered by deCODEme is that of analyzing the genome with respect to specific traits and health risks. For some weeks I resisted the lure of the health results. Both of my parents had struggled with cancer, and I was not that interested in the kind of fortune telling offered by personal genomics. I guess news of the New York spit party helped to change my mind. Somehow, collective spitting appealed to the anthropologist curious about the social life of DNA and the implications of the new genetics in the modern age. My results for the forty-seven diseases and traits covered at the time were based on calculations comparing my genome to sequences of participants in studies published in the scholarly literature. To access results for some diseases, I was invited to read about the genetic and medical details and to sign a statement about informed consent by clicking on "Accept." I need not bore the reader with the personal details. Suffice it to say that some of the information provided sounded trivial (no alcohol flush reaction), some of it resonated with what I thought I already knew (I am less likely than the general population "to become nicotine dependent [15 percent or less]"), some results were encouraging (I have low lifetime risk of some diseases, much less than for males of European ancestry in general), and some details might promote the hypochondriac in me to request further medical information (my risks for some diseases are slightly higher than those of my genetically significant others). When presented with these results, I was offered details on the mathematics of risk analysis. Also, I was invited to zoom in on my genomic landscape, focusing on a part of a chromosome and the location of specific mutations reportedly responsible for potential traits or diseases. Again, there were some surprises and some food for thought.

The Genome Browser of deCODEme allowed users to compare their complete data with friends and family. While my reference group of friends and family included both hypochondriacs and anthropologists, they had seen few good reasons to participate and, as a result, there was not much to compare. The website, however, allowed me to examine my genetic sharing with "famous" people. Here, sharing was indicated visually by the coloring of the relevant bits of the chromosomes. No doubt personal genomics is becoming both a family affair and a global concern. At any rate, a thriving imagined community of the users of personal genomics projects has been developing on the Internet.

2.4 Cyberspace: the experts and the rest

A number of websites testify to a lively discourse on the issues involved, including thinkgene.com, dna-forums.org, Eye on DNA, Urban Semiotics, and Dienekes' Anthropology Blog. The last one is dedicated to human population genetics, physical anthropology, archaeology, and history. Judging from these websites, there is more interest in exploring ancestry than health risks. Perhaps users are reluctant to reveal their health risks in public, although they may be keen to download the relevant information for their own purposes. Some of the websites referred to are focused on specific personal genomics projects while others are more general. Users engage with the goals of personal genomics, analyses of their own genome, and comments expressed through the expanding virtual community of the Internet.

Often, users comment on each other's roots and genetic identification, usually identifying themselves by first names or nicknames. In World Families Forum (2008), one user offered the following statement, referring to one of Sykes's seven daughters of Eve, Ursula: "Both my wife and I are U5a1a (Clan Ursula) members and on my Y side I am an R1b1c. We believe we are part of the Vandals, Visigoths, and the Normans groups of people who settled in Sicily – who were ancient Vikings/Scandinavians." Another user responded: "Some information about your R1b1c ... It is unlikely that it is of Viking descent. More likely it originated in western Russia." In some cases, users deliberately request advice or interpretation: "This is a nice website and I have enjoyed it. I have recently had my mtDNA tested and have been identified as a U5 (weak match). Can anyone please tell me what the 'weak match' means? I thought a match was a match. Thanks bunches!!"

Clearly, this is a biosocial community in the making, social networks based on identification with genomic characteristics. A certain "countess" commented, "I just found out that I am a daughter of Ulrike. Wow, how amazing to feel this connection to the past ... My ancestry is almost all British Isles, but I didn't know we were descended from Vikings!" (Family DNA 2008b). Another user, "PDHOTLEN," reported after "psyching" himself or herself to have the results of a mtDNA test: "My results (U5) say that my maternal ancestors were responsible for the demise of the Neanderthals in Europe ... Does anyone out there have a similar mtDNA?" Soon there was a response: "I also have U5, but my ancestors were probably not among the cave decorating Cro-Magnons. They are the Saami people up in the northwestern corner of Europe ... Getting the results from the genetic tests has made me a hobby genealogist and with good helpers I have been able to trace some Saami relatives back to the beginning of 1600" (Family Tree DNA 2008).

While many users become knowledgeable and skillful readers of genomic texts in the process of blogging, sometimes the technical jargon becomes overpowering, reaffirming the dividing line between experts and laypersons. Consider the following exchange.

DIDIER: The problem with the shorthand naming is that you lose contact with the subgrouping. Only those familiar with R1b haplogroups would know that R-U106 is distinct of RU152, R-SRY2627, R-L21 ... I would favor a simplified nomenclature ... The others are cryptic, very interesting for the specialist (those of us discussing these issues are specialists) but very confusing for non-specialists.

BANKS: It appears to me that you specialists will be preaching to the choir because the congregation (non-specialists) will be long lost in a state of utter confusion ... I have no knowledge of the "longhand" let alone the shorthand of DNA classifications. I only recently learned to spell haplotype. (World Families Forum 2008)

Occasionally, renowned experts drop in to offer a point of view or to clarify some issue. At one point it seems that James D. Watson, who shared a Nobel Prize for unraveling the structure of DNA, entered the scene, to underline the reliability of the testing provided by personal genomics: "It is important to note that no one is questioning the accuracy of the actual sequencing behind these services"; immediately, there was a respectful response from "docduke": "If that last post was by James D. Watson, I would like to take this opportunity (if you check back), to thank you for your candid, and honest, statements here as elsewhere. Intellectual honesty is becoming altogether too rare in public scientific discussions these days!" (Chron.Com 2008).

Quite possibly, the name "docduke" suggests that the blogger's point in referring to "candid, and honest, statements" is primarily to show support implicitly of Watson's statement in the *Times* of London about race, genetics, and the intelligence of many Africans: "All our social policies are based on the fact that their intelligence is the same as ours – whereas all the testing says not really." Because Watson's own genome is online, his ancestry is open for scrutiny. Exploring the data, the deCODE team concluded, with a hint of irony, that whatever Watson might think of black Africans, 16 percent of his own genes probably come from black ancestors (ABC News 2009).

Some bloggers take a playful attitude to genome testing. One woman who had her husband "tested" for fun was questioned about her ethics.

MEGAN: I admit it. I have no self-discipline when it comes to genetic genealogy. When deCODEme was launched, I had to be one of the first in line to get tested. So I ordered ... and received results ... my husband's results, that is. I thought this might be a little more interesting since he sports a Y chromosome.

MARIE: Did you get your husband's permission before using his DNA? ... How does he feel about you sharing his results with the world?

MEGAN: Yes, rest assured, I checked with him several times just to be sure. (RootsTelevision 2008)

Sometimes people check whether they are being cheated by genomic services. One blogger claimed to know of "at least one case ... where a customer deliberately submitted a dog's DNA just to 'test' the company. He was willing to pay for his little experiment, and yes, the company figured out exactly what had happened!" (RootsTelevision 2008). Sometimes users subscribe to two or more services to compare their usefulness and explore their methods and reliability.

I tested for both 23andMe and deCODEme. I just received my 23andMe results and I am quite surprised by the admixture test named Ancestry Painting. Indeed, I am half Berber, half French (E3b1b-M81, MtDnapI) but my 23andMe results showed 100% European (African and Asian p 0%) whereas my deCODEme results showed 81% European, 13% African ancestry and 6% Asian ... Do you know if these very different results can be because of different method used between 23andMe and deCODEme? (Dienekes' Anthropology 2008)

Concerned more with estimates of health risks than ancestry, Francis Collins, director of the US National Institutes of Health, caused a stir in 2009 by announcing that he had signed up for several personal genomics services under a false name, comparing and testing their prognoses. While sequence-wise, he concluded, the tests appeared to be "highly accurate," the final risk score varied from case to case: "[O]ne company used 5 single nucleotide polymorphisms, or SNPs, to calculate risk for a particular disease, pronouncing Collins at low risk. Another used 10 SNPs, placing him at high risk, and the third used 15, concluding that he is at average risk" (ScienceBlogs 2009).

It seems that the virtual community of genetic citizens is actively debating and negotiating roots, identities, and health risks by fusing the expertise of geneticists and nonprofessionals for the purpose of scrutinizing SNPs and comparing haplotypes. To some extent, however, the experts and the public engage in separate conversations. Nevertheless, representing just a tip of a rapidly expanding iceberg, the preceding excerpts from the Internet give some idea of the discursive community involved, the relations established, and the concerns people have.

2.5 From genome to identity

In some of his last writings, Foucault (1988) shifted his attention from systems of domination to the agency and experience of the individual, drawing attention to the particular kind of subjectivity characteristic for the modern age and what he called "technologies of the self." It seems reasonable to argue that personal genomics represent one example of technologies of the self

(Hacking 2006). Indeed, the genomics of ancestry is often assumed to provide an important avenue into identity and personhood. Significantly, Sykes's (2001) book on ancestry, which opens with the question "Where do I come from?," closes with a chapter titled "A Sense of Self." Anne Wojcicki, the cofounder of 23andMe, also underlines this point; the 600,000 genetic markers interpreted by 23andMe, she argues, are "the digital manifestation of you" (Hamilton 2008). Knowing where we come from, we apparently also know who we are.

Not only is there a growing popular literature that equates human beings with their genomes, this is also a theme explored in endless blogs on the websites mentioned above. One of the relevant statements suggests, in a somewhat humorous tone, that DNA "speaks" to its host:

> I received my DNA results earlier this year and was surprised to find myself in clan Ulrike. I have traced six generations of maternal ancestors in the Beds/Northants borders region. The Viking invaders did travel into this area . . . I have always been attracted to northern wilderness and have visited Alaska, Greenland/Iceland and Siberia. Is this my DNA speaking?! (Family DNA 2008a)

DNA, then, from the past tends to be seen as an avenue into the future, our essence and fate.

Battaglia (1995: 3) suggests an "approach to selfhood as embodied and historically situated practical knowledge," which in her view "prompts a larger question of rhetoric, namely, what use a particular notion of self has for someone or for some collectivity." Given such an approach, an important theme on the anthropological agenda is to explore the rhetoric of personal genomics and the pragmatic uses of genetic notions of self for the actors involved (see, for example, Gibbon and Novas 2007). Nowadays, with personal genomics, the collection and analysis of DNA are closely linked to commercial marketing of identity, including "race." Anxious to explore the unique signatures of their genomes and to care for their selves and their bodies, people are buying in rather than being solicited or tracked down. In the process, they facilitate the construction of gigantic DNA assemblies, coproducing knowledge of genomic differences.

2.6 Conclusions

As we have seen, there is a rapidly growing interest in personal genomics for the purpose of reconstructing our past and celebrating our emerging biosociality and for managing our lives and our future. Day by day, the companies involved have offered additional services on their websites, further details on diseases and traits, higher resolutions of data, and more powerful machines, diagnostic chips, visual presentations, and interactive features.

Anthropology is implicated in these developments at several levels, contributing to the key data sets employed on human history and variability.

Several personal genomics services provide information on both ancestry and health risks. For consumers, it seems, the former kind of information, the extraction of history from what Zuckerkandl called molecular "semantides" (see Sommer 2008: 506), usually involves a fair amount of playfulness, sometimes with undertones on identity, race, and networking, while the latter is inevitably associated with serious issues relating to medicine and lifestyle. Explorations of ancestry and health risks, however, overlap, enmeshing users in new biosocial relations and networks. One may wonder whether the relative importance of the two kinds of services may not be considerably altered in the future.

There are serious anthropological and philosophical doubts about important issues relating to ancestry, including the shape of the family tree, the validity of the molecular-clock hypothesis, and the sampling of populations (Bolnick *et al.* 2007). Analyses of ancestry in personal genomics services, however, are likely to remain more or less intact, partly because there is not so much at stake and, in any case, it is largely play. Studies of the genomics of diseases, in contrast, are riddled with contests, doubts, and conflict. Most common diseases, it seems, are only minimally explained by genetic factors, and in each case a great number of genes is likely to be involved. Last but not least, there is growing evidence for the importance of epigenetic factors beyond the simple concept of DNA sequence. Prainsack *et al.* express some of the doubts: "Personal-genomics customers are already going through a process of disenchantment: it is increasingly clear how little power SNP-based readouts of a person's 'genotype' offer for predicting future ailments in an individual" (2008: 35).

Given the evidence and the growing public awareness of it, why would people bother to consult their health risks with personal genomics services? It seems unlikely that the narcissistic pleasures involved in the exploration of ancestry and the genetics of health risks are withering away, given the central place of the human body in late modernity. Also, there are immense financial stakes and concerns on the global level for biotechnical and pharmaceutical companies. Moreover, the quality, magnitude, and comprehensiveness of knowledge can only increase with time. The power of computing machinery continues to expand, and cheap complete sequencing is within reach. As a result, one may expect personal genomics projects to expand, realigning experts and consumers, institutions and disciplines, including anthropology.

The narcissistic pleasures of late modernity are reaching new levels that neither Lasch (1979) nor Foucault (1988) could have anticipated. Personal genomics is just one example. Another example, perhaps, is the use of humans as model organisms in biological experimenting, as human cellular material replaces that of mice and fruit flies, a point emphasized by Rheinberger (2009: 8): "Many experiments are now being carried out with human cells directly ... With man becoming, in a sense, a model organism of his own, 'modeling' inevitably takes on the meaning and the form of human modification." One suspects this significant shift will have profound implications for anthropology and its understanding of the species and its variability, although the terrain is just beginning to be explored.

In recent years, the notion of the tree of life – a central notion for Darwin, Linnaeus, and much of biological thought since their time – has been seriously challenged by a strange mix of scholars such as W. Ford Doolittle, Stefan Helmreich, Tim Ingold, Marilyn Strathern, Gilles Deleuze, and Félix Guattari. For one thing, it has been reasoned, the tree of life might turn out to be a net or a rhizome, with endless reconnections rather than treelike bifurcation. While it is possible that theoretical challenges to the tree model and the empirical evidence available only minimally blur the main picture, simply shaking the tree a bit, they raise fundamental questions as to what should count as relatedness. Why, indeed, should we stick with reductionist gene talk? In the current age of biosociality, life has become increasingly disembodied, cultured, informatic, and rewritable.

Just as the history of physical anthropology and the bone collectors of the last century coalesced at a measuring party, the saga of modern biological anthropology and personal genomics coalesces at a spit party. In some respect, the two parties are rather different. Thus, whereas in the former case, research subjects were tracked down for measurements, often in the context of a colonial hierarchy, in the latter case, subjects offer themselves for the project in the hope that they may, in the process, discover and take care of their selves and bodies. This underlines radically different biosocial relations of production. The measuring party and the spit party, however, have one serious flaw in common – namely, the subtext of human variation, "race," and the presumed split between the biological and the social. After all, "biology" and "society" are not separate realities or categories of being. Human variation, including that identified as "race," is a thoroughly relational, biosocial state of affairs, collective history embodied in the habitus. Theodosius Dobzhansky once remarked that "nothing in biology makes sense except in the light of evolution" (Smocovitis 2012: 122). Given the deliberate conflation of the biological and the social in the wake of the new genetics, the human configuring of life itself, and the advancing evidence on epigenetics and developmental systems (Griffiths and Gray 1998), it would be more appropriate, twisting Dobzhansky a bit, to state that nothing in biology makes sense anyway, except in the light of the irreducible unity of the biological and the social.

The big challenge for anthropology now is to realign the biological and the social on new terms in a nonreductionist fashion. We can continue to craft our professional selves on two different tracks and to practice the study of humans as if it involved the investigation of two radically separated domains, defending the subdisciplinary boundaries as if they were engraved in our subjects, but it would be both ethnocentric and out of time. It is time to rethink the field on the assumption that *Homo sapiens* is an undivided being and that decoding it – to the extent that the language of "decoding" is the appropriate one – requires integrative perspectives that in the absence of a better non-dualistic language resonate with our biosocial natureculture. This will not be easy, but it is the only meaningful way to go.

The image and the report in the *New York Times* regarding the launch of 23andMe draw attention to the role of metaphors. Reporters quickly drew upon a series of related metaphors; the event was described as a "spit party," the message of 23andMe and personal genomics in general, it was argued, was "when in doubt, spit it out," personal medicine was said to be "within spitting distance," and so on. The people of 23andMe now have a blog site called "The Spittoon," drawing its name from an object also called "spitter," a receptacle for spitting into: "Using nothing more than a bit of saliva (Get it? The Spittoon!), the genotyping process we use analyzes more than 580,000 locations in a person's genome" (Spittoon 2009). It is tempting to assume that the spit is becoming one of the key metaphors we live by, informing our speech and our thoughts. Metaphors, however, just like DNA, frequently undergo mutations, recombining available material from everyday language and experience. The notion of the "spitting image," as we have seen, is a case in point.

The spit party seems to nicely capture various aspects of personal genomics and the gene talk on which it is based, the mechanisms of inheritance, the matching or mis-matching disclosed through the sequencing of DNA material, and the establishing of distance and ancestry, both genetic and social. When spitting out one's saliva one is presumed to provide a spitting image of oneself, encoded in DNA. The transparent metaphor has, finally, been elevated above the debates of etymologists. The emphasis, on the other hand, is no longer on spitting *at* someone (usually a gesture of contempt). Rather it is the conviviality of spitting *with* a fellow human being, for the purpose of celebrating biosocial bonds, for founding social networks based on bodily signatures. With a ginger martini.

3 Laboring lives: genomic stuff

Nowadays, the capacities of the fragmented body have been turned into instruments for production, redefining labor and agency. Not only do all kinds of agents and instruments labor *on* life, through the political economy of life itself, life itself *does* all kinds of labor. One example, I suggest, is genomic labor, involving the production of biosocial value which is material and informatic at the same time, both natural and social. This chapter extends the notion of relations of production beyond "natural" resources in the classical sense to the extraction, reproduction, and exchange of bodily material and information, to *biosocial* relations of production. While it may be argued, in a late-Foucauldian fashion, that the users of the genomics services discussed in the previous chapter perform labor *on themselves*, seeking to know themselves and to care for their bodies, their labor also needs to be situated in the biomedical mode of production involved. Addressing the realities of life itself in late-modern times in terms of labor processes and relations of production helps to characterize the different arrangements involved in the production and circulation of biosocial value.

Classic theories of political economy developed by Adam Smith, David Ricardo, Karl Marx, and many others assumed a laborer developing and applying his or her skills *in situ*, typically in an agricultural field or an industrial factory. For them, labor activities took place in the home or a nearby field or factory, in a local community embedded in a larger political economy. There have been exceptions, of course, throughout human history, with partial or temporary separations of the site of labor activities and the site of home. Nevertheless, in agrarian and industrial discourse the sites of labor and home have tended to be more or less permanently collapsed, a single life world or habitat.

In recent years, however, the spatial relations of labor activities have been radically changed as a result of complex and interrelated developments, including the growth of the World Wide Web, the network society (Hardt and Negri 2000), and virtual migration (Aneesh 2006); "paradoxically," Aneesh (2006: 2) notes, "the new space of transnational labour has reversed its relationship with the worker's body. Rather than move the body across enormous distances, new mechanisms allow it to stay put while moving vast quantities of data at the speed of light." Call centers of the kind studied by Aneesh underline the ability to perform work at a place other than the site of the acting body. What travels here is not the worker, but her voice and presence, in its digitalized form of bytes and megabytes.

Not only have the sites of labor and production increasingly been separated, the capacities of the body have been fragmented and turned into instruments for production, redefining both human labor and human bodies. Thus, the famous HeLa cell line taken from Henrietta Lacks who died from cancer in 1951 are reproduced on a global scale for a variety of purposes independent of the original host (see Landecker 2007). "Pluripotent" stem cells, endowed with the capacity to generate a variety of body tissue, represent another example. The "same" body, then, in a sense, performs labor at two or more sites simultaneously. Not surprisingly, these complex tournaments of bio-social value pose intriguing questions about biopolitics, place, and agency. A broadened notion of *biosocial value* is essential, I suggest, for understanding the coproduction of the hybrid complex of genomic material and information. Thus, the users of personal genomics services often contribute both a biological sample (a swap or a spit) and a variety of information on background and lifestyle. How do people become implicated in new labor processes as they subscribe to personal genomics services?

The companies involved in personal genomics have tended to emphasize consumers' access to medical knowledge and their relative independence of the medical establishment; hence, the claim on the deCODEme website: "[W]e wanted not only to empower the public, but also to give students, academics, physicians and other professions with an interest in genetics a chance to get a more in-depth view of their code and genome" (deCODEme 2009). Indeed, users draw their own conclusions and engage in dialogues with genomic experts, sometimes becoming experts themselves in the process. In a sense, then, this is science from below (Harding 2008). One example is SNPedia. Drawing upon summaries of peer-reviewed articles presumed to be relevant for given genomic data, it allows users of different testing services to pool personal data, to learn more about their own genotypes, and to explore the effects of variations in DNA. The consumption of personal genome data, as Prainsack (2011) points out, is also sold as an experience. Indeed, for me submitting samples, browsing the websites, downloading the results, and discussing them with family and colleagues was both an experience and an opportunity to look at potentially relevant medical information that my family doctor was unlikely to have access to. Interestingly, when I presented my results to my family doctor she appeared not to know of the deCODEme service. This is not, however, the end of the story. As we will see, one should not be too easily seduced by the rhetoric of democracy and of care of the self. It is important to draw attention to the laboring consumer.

Broadening the feminist perspective of Dickenson (2007) and some others, I suggest that the labor carried out by *both* men and women when subscribing to personal genomics services largely goes unrecognized. Also, this labor is both material and "immaterial," contributing personal information regarding lifestyle, diet, and so on as well as DNA material. Moreover, in contrast to the neoliberal "body shopping" represented by nannies and cleaners who physically move to the site where they are needed (Freeman 2011), the providers of genomic material and information are virtual migrants, in Aneesh's (2006) sense, at someone's service, contributing to transnational biobanks and databases that can be operated from anywhere anytime through the aid of the Internet and computing machinery.

Prior to the development of biotechnology and assisted reproduction, there was no place for human body parts in the Marxian scheme of the labor process although Marx did have something to say about *whole* bodies, estranged workers and slaves in particular. Thus, Marx refers to "nature" as "inorganic body; that is to say nature, *excluding the human body itself*" (1959: 126–127; emphasis added). Given Marx's framework, the dual identity of the human body as a laboring phenomenon *and* as an object of laboring activities is a contradiction in terms. While the extension of human mastery to the body itself complicates the Marxian scheme, it also invites intriguing questions about labor, production, and ethics. As Dickenson (2007: 29) emphasizes, "modern biotechnology muddies the clear distinction between things external to our bodily selves and those intrinsic to us . . .. The notion of 'external' is problematised and problematic in modern bioethics and biolaw, and with that come difficulties that Marx did not have to confront about what is alienable and what is inalienable from the subject."

Engels, however, suggested that one might think of the body itself as the product of labor. Drawing upon an evolutionary perspective, he argued that "in a sense, we have to say that labour created man himself" (Engels 2007: 25). "[T]he hand," he went on, "is not only the organ of labor, it is also the product of labor" (Engels 2007: 26). The modern world of biomedicine has made Engels's man-makes-himself perspective more pertinent than he could have imagined. Thanks to their labor activities, humans are now able to reproduce their own bodies, as part of the "inorganic body" of "nature." Not only do modern bioindustries produce a variety of "biologicals," agents extracted from or generated by biological material, these biologicals perform their own labor. Stem cells, for instance, are increasingly cultivated outside the human body, producing organs as "spare parts" for humans. Clearly, with modern biotechnology the "natural" capacities of the body have been turned into instruments for production, redefining both human labor and human bodies.

3.1 Producing bodies

Marxian theory is very much an agrarian discourse extended to industrial, capitalist production. Underlining human dependence on the environment (usually the land) through the production process, Marx (1959: 275) suggested that nature and humans formed a single "body": "The life of the species, both in man and in animals, consists physically in the fact that man (like the animal) lives on inorganic nature; and the more universal man (or the animal) is, the more universal is the sphere of inorganic nature on which he lives." In Marx's view, during the period of agriculture the land was "still recognized as a phenomenon of nature independent of man – not yet as capital, i.e., as an aspect of labor itself. Nature appears rather as an aspect of the land" (Marx 1959: 292). With the introduction of capitalism, nature was redefined as the expanding space for alienated labor, as an "aspect of labor itself." As Schmidt (1973: 82) puts it in *The Concept of Nature in Marx*, in agricultural production nature is "absolutely independent of men, men are abstractly identical with nature. They lapse, so to speak, into natural existence. However, where men succeed in universally mastering nature technically, economically and scientifically by transforming it into a world of machines, nature congeals into an abstract in-itself external to men."

Marx and several other commentators on the British "Factory system" discussed the key changes introduced by the mechanization of Victorian industries by means of prosthetic metaphors, in terms of relations between parts and wholes, organs and machines. In large-scale industrial production, Marx emphasized in *Capital*, each person was "bound hand and foot for life to a single specialized operation," a labor process that converted the worker into "a living appendage of the machine" (1976: 614). While responses varied – some being paranoid and others enthusiastic – different commentators addressed fundamental questions about identities, priorities, and hierarchies: "As prostheses are grafted upon yet other prostheses, which entity takes the role of attachment and which of host? In these hybrid couplings, what is actually a part of what?" (Ketabgian 1996: 13). What happens, then, when human mastery is turned inwards, extended to the bodies of the laborers themselves? Do they, perhaps, "lapse" once again into natural existence?

When discussing the production and reproduction of body parts, it is essential, of course, to pay attention to differences in the material in question. The "candidacy," in Appadurai's (1986) sense, of genes, cells, tissue, organs, and embryos for extraction, reproduction, commodification, and exchange is highly variable – depending on material properties, technologies of extraction, storage facilities, reproductive opportunities, and the cultural framing of human bodies – although candidacy has been extended to practically anything living. As a result, body parts bring "donors" and "targets" – persons and laboratories, people and markets, agencies and consumers – into different kinds of relations that ethnographers and theoreticians are busily exploring these days. Brief examples will have to suffice in this context. The general point to make is that extending the notion of relations of production to body parts – to the extraction, reproduction, and exchange of bodily material – may help to characterize the different arrangements involved in the production of biocapital, with their practices of ownership, relations of hierarchy, subjectivities, and sense of personhood and identity.

Human blood, to take one example, is routinely assembled and banked, for a variety of practical purposes, usually on a voluntary basis, within the framework of the gift. The Maussian gift, as Dickenson (2007: 21) puts it, "is still in a sense alive – far more so than even [Mauss] . . . might have realised, in the case of biological tissue" (see also Schneider 2003). Some human biological gifts are considered waste; human tissue samples are habitually extracted at hospitals for diagnostic testing and for material documentation of surgery. "Donors" are usually understood to have freely relinquished their samples, with or without informed consent, on the assumption that such tissue is of no practical use. New theoretical frameworks and advancing technology, however, may turn such "trivia" into gold mines.

Most extracted human biological material is neither waste nor a gift. Human reproductive material, in particular, is usually highly controversial, involving contested concepts, schemes, and relations. Thompson (2005: 248) suggests that a move beyond production characterizes reproduction centered on the human embryo, a mode of reproduction that "has its own characteristic systems of exchange and value, notions of the life course, epistemic norms, hegemonic political forms, security, and hierarchies and definitions of commodities and personhood." The practice of "surrogate" motherhood has received a fair amount of attention in both the media and academic works, partly because of the complications it invites for legal definitions of parentage and motherhood. To underline the human labor and biosocial relations involved, Dickenson (2007: 54) applies the notion of "biological lumpenproletariat" (originally coined by Dorothy Nelkin in another context) to women who rent out their wombs.

While in a sense sperm and ova, the key resources for assisted reproduction, represent identical components, each of them contributing essential genetic material, in practice they seem to represent radically different regimes of bodily commmodification. For instance, a study of egg agencies and sperm banks in the United States shows how "the dynamic interplay between biological, economic, cultural, and structural factors differentiates the market in egg from that in sperm in each stage of the donation process" (Almeling 2007: 336). In particular, different contracts are applied to egg and sperm donations with different kinds of rewards and etiquettes and different concerns for privacy. Overall, there is a mounting demand for human ova in biomedicine. This is due to both the growth in assisted reproduction in many contexts and the escalating demand for ova in the rapidly advancing stem cell industry. Thus, Hwang Woo Suk is reported to have used no fewer than 2,200 eggs from 129 women for his infamous stem cells project (see Gottweis and Triendl 2006). The shortage of ova is exacerbated by the fact that harvesting requires invasive surgery. Women's contributions not only invite a series of complications, including the risk of ovarian hyperstimulation syndrome, they are labor-intensive, productive work, not "merely" reproductive labor. As Dickenson (2007: 68) argues, battling with the patriarchal position of Marx who tended to treat women's domestic work as purely natural, not social, following an ancient, philosophical tradition, "it takes a great deal of intentionality and control to undergo the threefold processes of ova donation; of course it is labour, and hard labour at that. Women have a genuine Lockean property in the labour of ova extraction."

The market for human body parts is increasingly a global one. Both so-called reproductive tourism and international trafficking in organs point to a North–South division of labor, the people in the Global South generally offering cheap sources for the rest – as biological lumpenproletariat. Gibbon and Novas (2007: 13) argue that a "pressing issue for sociological and anthropological analysis is whose bodies are open to genetic and molecular remedies," pointing out that in some contexts (in India, for example) there is "great disparity over the social position of those whose bodies' embryos, cells and tissue are extracted, in contrast to the persons into whom they are inserted." Clearly, the North–South division underlines particular kinds of biosocial relations of production.

The human body, it needs to be emphasized, is not only being refashioned through the fragmenting, reshuffling, and hierarchies that characterize biosocial relations of production. Taking a broader perspective, it is also the object of a rapidly expanding production system – a global body shop, if you like – involving the manufacture and marketing of drugs and food products, including genetically modified food. Here, perhaps, Marx's later works on the "structures of capital" might be more relevant than his early works on alienation and estrangement. Increasingly, the activities of multinational corporations affect the *environment* from which human bodies receive their energy and nourishment no less than their internal constitution. What Marx would refer to as "man's metabolism with nature" is saturated by biopower of one form or another. This is one aspect of the New Economy of Late Capitalism (Fisher and Downey 2006).

The preceding discussion indicates that the trafficking in body parts is highly gendered, women's business. The tide, however, is turning. Dickenson (2007: 32) reasons that with the "new enclosures" of the human body as a result of advances in biotechnology "some aspects of objectification which were previously limited to women's historical experience are now being extended to biologically male bodies as well." Perhaps this is why there is renewed interest in early-Marxian notions of alienation and estrangement. The prospects of objectification concern everyone.

3.2 The organic and the inorganic

In the process of requesting and using personal genomics services, consumers facilitate, either directly or indirectly, the construction of gigantic DNA assemblies, coproducing knowledge of genomic differences. The spokespersons for 23andMe, unlike those of most of the other projects, were quite open about the issue of alternative uses of their data – for example, research. Wojcicki suggested that signing up for 23andMe was "a great way for individuals to be involved in the research world . . . You'll have a profile, and something almost like a ribbon marking participation in these different research papers. It will be like, How many *Nature* articles have you been part of?" (Goetz 2007). This is highlighted in a comment on one of the websites: "If the company succeeds in attracting the hundreds of thousands of customers Avey and Wojcicki talk about drawing, it will be sitting on one of the largest genetic databases on Earth. And there's no opting out of any research studies 23andMe wants to conduct at that point, either, since the consent forms to which customers must agree specifically commit their genome-scan results to future research" (Hamilton 2007). Arguably, the people contributing cheek swabs to personal genomics services are engaging in co-working, a collective labor process that ultimately results in large-scale biobanking. Spitting saliva and providing cheek swabs, after all, is biosocial work, potentially contributing to the global networks and hierarchies involved in the manufacture of biosocial value.

deCODEme was part and parcel of its mother company deCODE genetics, whose purpose is to advance biomedical research and pharmaceutical development. Although the company seems to have no plans to directly draw upon its personal genomics data in its biomedical research, a closer integration might take place later on. There are also strong financial and technical links between 23andMe and the giant Google which may be indicative of new, hybrid forms of biobanking and bioinformatics. Whatever their current ambitions, personal genomics projects are likely to connect with larger biomedical projects in the future. Given the possibility of hacking genomic data (Aldhous and Reilly 2009), the clients of personal genomics companies may eventually be contributing to projects beyond the awareness and control of the services they have contracted.

Perhaps the social network of Facebook helps to illuminate the issue of labor and co-working in the context of personal genomics. Here, users' expressions of "likes" with respect to particular retailers or services are routinely translated into advertisements, as a result of which Facebook collects profits from retailers' payments. How could users' work be acknowledged? A lawsuit filed in California in 2011 (ANGEL FRALEY, et al. Plaintiffs, v. FACEBOOK, INC., Defendant) provides some clues. The plaintiffs argued that Facebook users were not sufficiently informed of how their "likes" translated into profits. In response, Facebook proposed a settlement which would involve informing users about sponsored stories and a payment of $10 million to research and advocacy groups that work on digital privacy rights. A federal judge in California, however, rejected Facebook's settlement offer, requesting clarification on what would count as adequate and fair in such calculations and suggesting that lawyers might have "bargained away something of value" (United States District Court for the Northern District of California San Francisco 2012). Presumably the judge was referring to the co-working of users.

Hardt and Negri (2000: 30) emphasize the vital dimensions of biopolitics, focusing on the production and reproduction of life itself; for them, it would be misleading "to treat the new labouring practices in biopolitical society *only* in their intellectual and incorporeal aspects. The productivity of bodies and the value of affect . . . are absolutely central in this context." Perhaps it is necessary to react to the informatic, textual trend associated with the mapping of genomes and the "code of life." Indeed, life itself – in the form of stem cells, tissue, and organs – is a central component in the production of biovalue. Often, however, it is difficult to maintain a rigid distinction between the corporeal and the incorporeal.

This is particularly relevant for current gene discourse. As Keller (2000) famously argued, the concept of gene is highly unstable, and varies from one discipline to another. For Rheinberger (2000b: 221), similarly, the gene belongs to a class of fuzzy "objects" that cannot be assigned a precise meaning; in his view, the usefulness of boundary objects does not rest with a clear definition from the outset: "[I]ndeed it can be rather counterproductive, to try to sharpen the conceptual boundaries of vaguely bounded research objects while in operation." Keeping in mind the fact that the genome is sometimes regarded as informatic assembly and sometimes as a material thing, it seems to make sense, in order to avoid unnecessary ambiguity, to simply speak of "genomic stuff" (Palsson and Prainsack 2011); in other words, leaving aside the issue of materiality versus meaning. The notion of "biovalue," then, seems too restrictive when dealing with genomic stuff. It is easy, perhaps, to dismiss the notion of "stuff" in this context as it is often used in a pejorative sense to designate relatively worthless material or immaterial things, much like the notion of "junk." However, "stuff" is sometimes used with a very different meaning, conveying a heightened sense of vital importance. Thus, the

Middle-English term "stuffe," from which it is drawn, referred to both a person's essential moveable household property and the weapons and food necessary for battle (Harris 2011: 162).

To capture the complex implications of personal genomics services it seems pertinent to draw upon the notions of biosocial value and ensembles of biosocial relations. Users of these services are often contributing both tissue and information on phenotypic characteristics, health, drug use, and lifestyle – information that in the future will probably be updated interactively to increase the efficiency of the machinery of schemes such as deCODEme and 23andMe.

Prainsack (2011) suggests that if we are witnessing a "participatory turn" in genomics we need to ask what it involves. She makes a distinction between "early adopters" and "regular consumers" of personal genomics services, to highlight their different motivation and work. While early adopters may play a significant role in shaping the project, regular consumers are likely to be rather mute or passive. Contributors to genomic services and biobanks are not only doing work in the sense that they provide genomic stuff of critical importance; also they are engaged in the coproduction of biosocial networks. This is an issue highlighted in several recent works (see, for example, Thacker 2005, Levina 2010). Drawing upon her work on 23andMe, Levina (2010: 2) suggests that "life in the network society requires of its denizens a constant contribution to the growth of the network ... Members are encouraged to think of themselves as dividuals, or nodes, in the network." If network subjectivity, she continues, "is conceived in terms of dividual bodies and identities, then each body – reduced to its information – can be abstracted from its social and cultural context. It becomes, in a sense, a free-floating signifier" (Levina 2010: 7). Such a perspective seems to resonate nicely with the notion of humans as ensembles of biosocial relations.

3.3 Conclusions

No doubt, personal genomics of the kind discussed here involve an element of empowerment. Some qualifications, however, are needed. Prainsack *et al.* (2008) argue that while relaxing the genetic protectionism rampant in recent decades may be a good thing, giving people an opportunity to become active governors of their genomes, the arguments about individual freedom, informed choices, and the unregulated genomic marketplace should be taken with a grain of salt. For one thing, they disguise the fact "that personal genomics is pushing the individualization of responsibilities one step further" (Prainsack *et al.* 2008: 34). Anthropology and related fields can play an important role on this front by exploring what such individualization means and what people expect from genomics, providing "thick" descriptions (see, for instance, Santos *et al.* 2009).

Another qualification relating to the agency of the users of genomics services is also essential; this concerns the labor they perform *for* personal genomics services rather than their opportunity to comment, interpret, and engage in a dialogue on methods and results. Historically, the discourse on labor, property rights, and "resource" governance has described the characteristics of the regimes in question in terms of rather simple binary dimensions: stationary versus mobile, aquatic versus terrestrial, biological versus physical, material versus intellectual. Along with some other body issues, including surrogate motherhood, organ transfer, and biobanking (Dickenson 2007, Gottweis and Peterson 2008), genomic stuff seems to invite new dimensions and considerations. For one thing, with the new genetics, the development of biomedicine, and the expanding production of biocapital (Palsson 2007, Lock and Nguyen 2010), the very notion of the "biological world" has been destabilized as nature is increasingly subject to artificial, human refashioning.

Personal genomics, as we have seen, represents a series of developments in genomics, biomedicine, informatics, and neoliberal economies. Rooted in the personalized medicine characteristic for population biobanks founded during the past decade or so, aiming to produce medicine geared to individual genomes, it is now firmly embedded in network society. In some respects, the focus on labor in this context may be narrow and restrictive, economizing complex and diverse developments. Consumers, after all, seem to get some rewards, in terms of belonging and sociality, indulging in play and recreation. Also, one may argue, there is a crucial difference between reproductive and genomic laborers in that the former, unlike the latter, take serious health risks with considerable emotional and physical involvement.

Consumers of such services, however, I have argued, are not simply passive recipients of goods; they are actively co-working, contributing labor from different virtual sites at a variety of scales and hierarchies with different forms of alienation and exploitation. These biosocial relations need to be mapped in detail in comparative contexts, partly for their own sake to better understand the scene and partly with respect to policy and governance. Following Toffler who launched the notion of prosumption, Ritzer and Jurgenson (2010: 17) suggest that social theorists of production and consumption (for example, Marx and Baudrillard) have too strongly distinguished between these two spheres, suffering from either a productivist or a consumptionist bias. While "prosumer society," they argue, is nothing new, "a series of recent social changes, especially those associated with the internet and Web 2.0 (briefly, the user-generated web, e.g. Facebook, YouTube, Twitter), have given it ... greater centrality" (Ritzer and Jurgenson 2010: 14).

There is a vast terrain to explore in personal genomics and related fields, it seems, through the analytical lens of labor. The contribution of the "consumer" is negated twice, rhetorically and financially, in the fact that the consumer *pays for* contributing and *has no share* in the profits derived from patents claims and other potentially collective products of personal genomics. Recognizing the roles and complexities of co-working and the immaterial labor involved, in the communal spirit of solidarity that increasingly characterizes bioethical discussions (Prainsack and Buyx 2012), it seems likely that new forms of management and public engagement will be explored for personal genomics. Perhaps in the near future governmental and non-governmental agencies along with private companies and users' organizations will experiment with some forms of DNA co-operatives, moving not only from *me* medicine to *we* medicine, to paraphrase Dickenson (2013), but also, to draw upon the fashion of naming in the industry, from LaboringMe to LaboringUs.

For Marx, a series of concepts served to underline the worker's loss of control and self under the labor conditions of capitalism: estrangement, embodiment, and externalization (*Entfremdung*, *Verkörperung*, and *Entäusserung*, in German). Given the relations of production involved, the terms of trade associated with private ownership of the means of production, the worker's labor appears as something external to him- or herself, as belonging to someone else: "[T]he external character of labour for the worker appears in the fact that it is not his own, but someone's else's, that it does not belong to him, that in it he belongs, not to himself, but to another" (Marx 1959: 274). The external character of labor and the loss of self imply that the end products of laboring activities, the goods embodying a part of the worker, are separated from the person: "The *alienation* of the worker in his product means not only that his labour becomes an object, an *external* existence, but that it exists *outside him*, independently, as something alien to him, and that it becomes a power on its own confronting him" (Marx 1959: 272; emphasis in the original). Rather than affirming themselves in their work, Marx suggested, workers *deny* themselves, undermining their integrity, agency, and personhood. Human labor, in other words, is being reduced to an "inorganic" condition, much like the labor of the "natural beings" of the agrarian project, including slaves and cattle, usually regarded as appendages of the earth.

To fully understand the implications of the Marxian concept of estrangement and its relevance for the analysis of the current fragmenting and realignment of bodies, it is pertinent to attend to the idea of species-being. For Marx, each species has its own species-being manifested through its engagement with the environment, reflexivity being the key character of *homo sapiens*: "The whole character of a species – its species-character – is contained in the character of its life activity; and free, conscious activity is man's species-character. Life itself appears only as a *means to life*" (Marx 1959: 276; emphasis in the original). One of the key paragraphs in the *Economic and Philosophical Manuscripts of 1884*, which is worth quoting at some length, elaborates on the idea of "character" by underlining Marx's pragmatist take on human consciousness and its firm grounding in the "life-activity" of the real world:

> It is just in his work upon the objective world . . . that man really proves himself to be a *species-being*. This production is his active species-life. Through this production, nature appears as *his* work and his reality. The object of labour is, therefore, the *objectification of man's species-life*; for he duplicates himself not only, as in consciousness, intellectually, but also actively, in reality, and therefore he sees himself in a world that he has created. (Marx 1959: 277; emphasis in the original)

Emphasizing the species boundary and the uniqueness of *Homo sapiens*, Marx's anthropocentrism is rather problematic; modern readers are unlikely to restrict agency to humans the way Marx did. For Marx, however, the contrast with non-human animals was useful in hammering home the central point about estrangement and the associated reshuffling of nature and society under particular conditions of production. "In tearing away from man the object of his production," therefore, Marx (1959: 277) went on, "estranged labour tears from him his species-life, his real objectivity as a member of the species, and transforms his advantage over animals into the disadvantage that his inorganic body, nature, is taken away from him." It is precisely here, as several authors have pointed out, that early Marx becomes particularly pertinent for analyses of the condition of the biosocial. The estrangements represented by the biotech mode of production – the fragmenting, trafficking, and hybridity of body parts and the biosocial relations in which they are embedded – mirror, up to a point, the objectification and alienation of "species-life" addressed by Marx. As Thacker (2005: 40) puts it, "Marx's species being is transformed into a 'molecular species being,' a species being in which labor power is cellular, enzygmatic, and genetic."

The alienation of persons from their body parts takes many forms, depending on a host of factors. The *bio-graphies* of fragmented bodies, in the literal sense, their life-courses, unfold through the agency of a series of actors and actants, in the Latourian sense, who both constitute and are constituted by particular biosocial relations of production. Thacker (2005: 308) suggests that fields such as tissue engineering offer "the prime context in which to investigate the extent to which our own bodies – as biologically constituted – are deployed as actants . . . The tissues are my own, yet they exist outside of and separate from my body." As Lock (2007: 225) has shown through her ethnographic work on organ donation, "organs very often represent much more than mere biological body parts; the life with which they are animated is experienced by recipients as personified, an agency that manifests itself in some surprising ways and profoundly influences subjectivity."

It may be tempting to read Marx as an essentialist, assuming preformed character for each species and every organism along the lines of current gene-centrisms. While the notion of species-being clearly introduces some kind of biology into political economy, it is a relational notion emphasizing the coproduction of the organism and its "inorganic body," the environment. Marxian theory on this score, therefore, can easily be reconciled with theories underlining mutual interactions in the constitution of life.

4 What's in a genome? Indigenous encounters

By rendering obsolete the theoretical opposition of nature and culture, the study of the human genome has given rise to fresh networks among anthropologists and other scholars. These developments, in turn, invite a refashioning of anthropology. While anthropology has often been divided on the issue of genomic research (Marks 2001, Lock 2005), with the conflation of nature and culture in the wake of assisted-reproduction technologies (Strathern 1990), molecular genetics (Rabinow 1996), and the conceptual, technical, and institutional developments associated with the recent Human Genome Project, a fresh network of associations has been emerging among anthropologists studying the human genome. These developments invite interesting and pressing questions: How do analyses and conclusions about genomic variation and history generated through laboratory work relate to local notions of subjectivity, belonging, and relatedness? What kind of relations between researchers and the people providing ethnographic and genetic information (formerly "informants" or "sample donors") should one opt for?

Comparative research on the human genome has a complex history. In the early 1990s, the Human Genome Diversity Project (HGDP) was organized to explore human differences and history by sampling fragments of the genome from a number of populations across the globe. An early critic of the HGDP was Donna Haraway. While she remained "sympathetic to the desire to produce a human species database that draws from as large a concept of humanity as possible," she suggested that the project's notions of difference, representation, and agency got "its well-meaning organizers into well-deserved trouble" (Haraway 1996: 353). To Haraway (1996: 353), the main problem was the absence of any formative role or partnership for the indigenous communities: "The people to be sampled ... were not regarded as partners in knowledge production who might have ends and meanings of their own in such an undertaking." During the more recent El Dorado scandal, the anthropological community was forced to take direct measures to avoid similar troubles in the future.

Here I use the term "genomic anthropology" to denote a hybrid field that increasingly involves anthropologists in the fusion of the "social" and the "biological" in the wake of genomic studies. Unfortunately, "biology" has colonized and narrowed the broad Greek notion of *bios*. Crawford's (2007: 79) characterization of "anthropological genetics" (in contrast to "human genetics") in terms of "its emphasis on smaller, reproductively isolated, non-Western populations" similarly seems too narrow and archaic, given the variety of roles that anthropologists are playing in genomic fields. Extending arguments originally developed by critical ethnography and feminist studies with respect to ethnography (see Lassiter 2005), I suggest that it is important to explore the implications of collaborative genomic research in which subjects become consultants and even co-researchers. Such an approach has its own problems, but it may allow for responsible practice and sensitivity to local concepts and relations, both of which are central for postcolonial anthropology.

Research in the Arctic – the northern rim of eastern Siberia, Alaska, the Canadian North, and Greenland – illuminates the possibility and the complexity of exploring the genomic trajectories of the past in a collaborative fashion while also drawing on other approaches, including archaeology, linguistics, and ethnography (Bellwood 2001, Jones 2003, Stern and Stevenson 2006). My discussion is partly framed by anthropological research among the Inuit, especially the Inuit Genetic History Project (IGHP) that I organized with Agnar Helgason (see Helgason *et al.* 2006). In a collaborative approach that conflates nature with culture and expert with lay knowledge, a central challenge for genomic anthropology is to engage with the intellectual theories of the people encountered in the field.

Until recently, the prehistory of the Inuit (a brief chapter in the long and complex history of the species) has been an underexplored area in research on human genetics and variation. Thanks to advances and collaboration in several fields, the charting of this history is now a realistic goal. Although on their own genes may not determine who we are or what happens to us, historical information on what Oyama (2000: 161) refers to as "developmental state" is necessarily embodied in the genome: "Developmental state is a kind of temporal slice through the life cycle. It carries the evidence of past gene transcriptions, mechanical influences inside and outside the organism, results of past activities, nutrition or the lack of it, and so on." The study of Inuit history can be significantly advanced through genomic analyses complementing and qualifying analyses of material remains, myth, and languages. While it is useful to explore the extent to which changes in genome, culture, and language go together, "the possibility of any 1:1:1 correlation between a gene pool, a culture, and a language, each changing only by internal variation of inherited source materials, can be dismissed right from the start" (Bellwood 2001: 201).

4.1 Inuit contexts

The early "zoning" of the earth into regions or cultures underlined Western ideas about cultural differences and the exotic. Thus, the "Arctic" (along with the "Tropics") often served as a mirror for the temperate "West," highlighting differences in climate, vegetation, and health. The Arctic, indeed, has been an important site for alterity (Hastrup 2007). The heroic search for the elusive Northwest Passage between the North Atlantic and the Pacific captured the imagination of Europeans for centuries. Occasionally, Inuit travelers and explorers have returned the gaze from a position radically different from those of Europeans in the emerging world system, with its hierarchy of identities, nations, and populations. Capitalizing on the popular imagination, by the end of the nineteenth century anthropologists and explorers collaborated in turning Eskimo culture into a museum piece for the pleasure and curiosity of the Euro-American public. Sometimes the Inuit themselves served as living specimens, one example being the Colonial Exhibit in Copenhagen in 1905.

At the beginning of the twentieth century, after a series of ethnographic expeditions, including those of Boas, Knut Rasmussen, and Vilhjalmur Stefansson, the Inuit became a staple of anthropological discussion. The physical and biological anthropology of the Arctic was also a twentieth-century development. While the first anthropological description of a Greenlandic skull by J.B. Winslow in 1722 generated speculation about Inuit origins and history, systematic physical measurements developed only during the first half of the twentieth century. Followers of the American tradition of Boas and Hrdlička showed "an almost fanatic devotion to measuring skulls" (Scott *et al.* 2000: 347), publishing detailed catalogs with thousands of measurements. With the modern synthetic theory of evolution that took shape in the 1940s, there was a sharp break with past emphases in biological anthropological research, in the Arctic as elsewhere, from physical measurements and race to studies of gene frequencies and populations.

Early debates on Inuit origins, migrations, and history were highly speculative, drawing on limited linguistic, mythological, and archaeological data and unreliable methods of dating. Steensby suggested in 1917 that the "dispute is old, and can still be said to be far from being finally decided" (1917: 41). By then, attempts had been made to establish Inuit origins on no fewer than three continents: Europe, Asia, and North America. The thesis of European origins rested on discoveries in the river basins and caves of northern France of Paleolithic bone implements that called to mind Eskimo artefacts. Writing in 1910, Boas discounted the European and Asian theses, opting for the Hudson Bay area of North America: "The much-discussed theory of the Asiatic origins of the Eskimo must be entirely abandoned. The investigations of the Jessup North Pacific Expedition [1897–1902], which it was my privilege to conduct, seem to show that the Eskimo must be considered as, comparatively speaking, new arrivals in Alaska, which they reached coming from the east" (Boas 1940: 337).

Although based on the results of an important expedition, Boas's thesis was embarrassingly off the mark. The broad outlines of early human migration in the Arctic have now been fairly firmly established; Damas's (1984) landmark handbook on the Arctic assembles the available evidence. More recently, a systematic synthesis of archaeological, linguistic, and genetic results has emerged (see, for example, Ames and Maschner 1999 and Fortescue 1997 for the Northwest Coast and Greenland, respectively). Doug Jones points out, partly with reference to the Arctic, that during the past decade or two an "emerging synthesis" based on correlations in the distributions of genetic, linguistic, and archaeological variation has gained increasing popularity in the study of prehistory. Thus, "Eskimo-Aleuts and Northern Na Dene are clearly separate from the remainder of New World populations, with distinctive gene frequencies and language families" (Jones 2003: 510). This approach echoes a general trend emphasizing that "increased knowledge and understanding will only come from careful multidisciplinary considerations of many strands of evidence ... This observation applies to archaeologists, linguists, paleoanthropologists and geneticists alike" (Bellwood 2001: 201).

Early theorizing on human settlement in the North American Arctic often assumed that the Inuit had lived unspoiled from time immemorial in isolated groups in practically the same location. Recently, a different narrative emphasizing the historicity and complexity of Inuit contexts has been gaining force. Throughout much of their history, the Inuit seem to have been dispersed in numerous, flexible, and mobile "*miut*" groups (that is, "the people of," for example, Killirmiut, Nunavummiut). It was only in the wake of the indigenous movements during the 1970s and 1980s that the people we now identify with the monolithic labels "Inuit" and "Eskimo" began to see themselves as a "people" or "nation."

The prehistory of human settlement of the North American Arctic is highly complex. Maxwell (1980: 161) suggested of his thorough attempt to provide an overview that "the temerity of such an enterprise borders on the ludicrous." Archaeologists speak of a series of populations and prehistoric traditions in the Arctic region of North America. The so-called Palaeo-Eskimos (Old Eskimos) seem to have reached the coast of Alaska from Siberia almost 5,000 years ago, according to radiocarbon dating. They probably represent the ancestors of the people referred to as "Tuniit" in Inuit oral tradition, the people encountered by the Inuit as they arrived in Arctic Canada. Some of their descendants, between 500 BC and AD 1500, developed the Dorset culture, named after the community of Cape Dorset, where their settlements were discovered. The other major population movement into the North American Arctic, that of the

"Neo-Eskimos," occurred around 1,000 years ago and reached the eastern Arctic by about 1150. The Neo-Eskimos adapted to aquatic resources with a social organization more complex than that of the Palaeo-Eskimos. Their early descendants in the eastern Arctic are often identified as Thule Inuit. Meanwhile, in the ninth century, after the settlement of Iceland, the Norse established colonies in southwest Greenland. With the eastward movement of the Inuit and the westward movement of the Norse, the two groups eventually met, probably in the eleventh and twelfth centuries. Encounters were described in both Icelandic sagas and Inuit narratives. After more than four centuries, the Norse colonies collapsed, apparently partly due to climatic reasons.

The modern Inuit of northern Alaska, Arctic Canada, and Greenland are frequently assumed to be the direct descendants of the Neo-Eskimos. An alternative hypothesis, however, suggests that the modern Inuit have descended in part from Palaeo-Eskimos. This hypothesis gained political momentum during the ethnic revival of the twentieth century, partly because ancient occupancy seemed to strengthen the Inuit case in negotiations of land claims. McGhee (2006: 113) suggests that

> archaeologists found themselves in a situation where academic interpretations were given political significance, and again it has become increasingly clear that the version of history advanced by those with a political interest was likely incorrect. Rather than being ancestral Inuit, it seems probable that the Palaeo-Eskimos can be identified with the people known to Inuit tradition as "Tuniit."

There are grounds for taking ethnopolitical rhetoric with a grain of salt, but McGhee's assertion that Inuit claims of Palaeo-Eskimo ancestry are "incorrect" may be an overstatement.

4.2 The Silk Road of the Arctic

While the Human Genome Diversity Project gave rise to intense anthropological debates about method, politics, and ethics, work continued. None of the fifty-two groups represented by the Centre d'Etude du Polymorphisme Humain (CEPH) panel, with the possible exception of the Siberian Yakut, are located in the Arctic. Genetic studies estimating the age of maternal (mitochondrial DNA [mtDNA]) and paternal (Y chromosome) lineages suggest an initial entry time of ancestral Native Americans of between 20,000 and 15,000 years BP (Schurr 2004: 571). The last significant phase of expansion into the New World included the Eskimo-Aleuts and the Na Dene. The archaeologically inferred mutation rate of the mtDNA control region often used to calibrate divergence into units of time in studies of human mtDNA sequence variation – the somewhat speculative rate of one transition per 20,180 years – was originally established in archaeological research in the

Arctic and was based on the premise that variation observed among a few Eskimo and Na Dene sequences had accumulated over a period of 11,300 years (Forster *et al.* 1996).

The outlines of human migrations into the North American Arctic are rather well established, but their details are still matters of debate. In particular, the relationship between the late traditions of the Dorset people and those of the earliest Neo-Eskimo Thule Inuit has remained unclear. There is still some truth to Maxwell's (1980: 170) statement that "termination of Dorset culture and its relationship to Thule encroachment from the west remains a paramount problem." It seems increasingly likely, however, that the Dorset and the Thule Inuit coexisted for a period and were aware of each other. Mary-Rousseliére (1984: 443) suggests that remnant Dorset survivors (Sallirmiut) on Southampton Island and at Foxe Basin were known to the Inuit; the Sallirmiut "were considered to be barbarians, and no one maintained relations with them." Recently documented oral histories from Canadian Inuit describe early encounters between Inuit and Tuniit. These accounts tend to dwell more on the material remains of the Tuniit and their conflicts with Inuit than on what they might physically represent.

The archaeological debate on Dorset and Inuit settlements during the period from 800 to 1150 seems to be tilting toward the account from Inuit oral history – that is, the idea of temporary coexistence. Basing his research in an area known as Iqaluktuuq, near Cambridge Bay on Victoria Island, Friesen (2004) presented new excavations and radiocarbon dates. To him, the Iqaluktuuq area is a regional archaeological hot spot, with a fairly complete archaeological sequence for much of the central Arctic. One reason is that the Ekalluk River, along which the key sites are located, served to concentrate the vital warm-season resources of arctic char and caribou. Friesen's results indicate that the Late Dorset lived in relatively close proximity to Thule Inuit for at least a century (2004: 689). Multiple dates from his three Late Dorset sites suggest relatively recent occupation, during the twelfth century (firmly within the period of early Thule Inuit occupation), indicating some kind of contact. Victoria Island was the center of an early archaeological debate involving the possible intermixing of Norse with Inuit and Dorset. Recent archaeological findings indicate that occasional contact occurred between Norse and Dorset in the Canadian eastern Arctic, possibly over several centuries.

While the archaeological and linguistic records resolve some riddles, they cannot tell us whether Dorset-Inuit, Inuit-Norse, or Inuit-Dorset contemporaneity involved exchange of goods, population replacement, or interbreeding. One avenue for resolving the issue is DNA analyses of ancient bones; another is genomic anthropology among living Inuit. Agnar Helgason and I took the latter route, launching the Inuit Genetic History Project, a research project on the genetic history and migrations of Inuit populations in the North American Arctic. We decided to base our project in Greenland and the Kitikmeot region of Nunavut. The Kitikmeot region seemed a good choice because of both recent reflections on Dorset-Inuit interactions and early debates about the possibility of contact between the Copper Inuit on Victoria Island and the Norse. The Kitikmeot area is quite possibly a genomic as well as an archaeological hot spot, the Silk Road of the Arctic. The project collected buccal swab samples from 299 people of presumed Inuit ancestry in the Kitikmeot region and fourteen locations throughout Greenland. The first publication based on the project (Helgason *et al.* 2006) examines mtDNA control-region sequences from these locations and compares them with previously reported sequences from Greenland Inuit, Chukchi, Siberian Yupiit, and Alaskan Aleut.

Basic results of this work include the following: (1) Greenland and Kitikmeot Inuit are relatively homogeneous genetically when compared with Yupiit, Chukchi, and Aleut populations; (2) significant differences appear among the subgroups of Greenland and between the groups of Kitikmeot, Siberia, and Greenland; and (3) there is no indication in the Kitikmeot data of Inuit-Norse admixture. Given the concern with the Dorset-Inuit riddle, these results are significant in that they seem to suggest "a complicated pattern of regional stratification in the distribution of individual haplotypes that is not easily accounted for by the prevailing hypothesis that all Inuit populations in Canada and Greenland are exclusively descended from Thule ancestors that expanded from Alaska only 800–1000 years ago" (Helgason *et al.* 2006: 132). The alternative scenario suggesting that expanding Inuit groups encountered Dorset populations in Canada and Greenland and interbred with them would, however, account for contemporary patterns of mtDNA variation in Canada and Greenland. This may indicate "a matrilineal contribution to contemporary Inuit populations of the Dorset peoples who inhabited Greenland and the Canadian Arctic prior to the Thule expansion" (Helgason *et al.* 2006: 132). Given this result and Friesen's recent dating from archaeological sites in the Kitikmeot region, it is likely that the Late Dorset and Thule people not only coexisted for a time but actually interbred. While these are preliminary results and more work is needed, the Inuit Genetic History Project illustrates the usefulness of genomic anthropology in exploring – along with archaeology, linguistics, and folk narratives – the trajectories of the past.

In many contexts ancestry is a highly sensitive matter, with potentially grave consequences for cultural identity, citizenship, and access to important resources. Anthropological research in this domain, therefore, must be particularly sensitive to local context and community concerns.

4.3 The IGHP: a brief ethnography

Mindful of the unpleasant history of research on human remains and genetic characteristics among indigenous groups and the marginalization of Inuit during the colonization of the Arctic, Inuit authorities have been anxious to monitor research in their jurisdiction. Kral and Idlout (2006) have outlined the changing Canadian context in the wake of the creation of the Nunavut Territory in 1999. Before the IGHP could begin, complex protocol procedures had to be followed. Formal applications were sent to Nunavut and Greenland in 2002. The goals of the project, as well as procedures of sampling and consent, were outlined in English, local Inuit languages, and Danish. The "experimental protocol" explained that the researchers' aim was twofold: "to examine the population history of, and genetic relationships between, Inuit groups" and "to establish whether there was any genetic evidence for early Norse admixture in Inuit groups from Greenland and Victoria Island in Canada . . . The data gathered for this study will . . . be used to explore the genetic history of Inuit populations in Greenland, Canada, and the Arctic, more generally – a task that has barely begun." The research protocol also had to be screened by the National Bioethics Committee of Iceland.

In order to seek permission to collect samples in Nunavut and to explain the project in person to scientific staff and community leaders, I visited both the Nunavut Research Institute in Iqaluit and the community of Cambridge Bay. In Cambridge Bay, meetings were organized with the elders of the community (the informal leaders) and the hamlet council. Some people suggested that the Inuit were "overstudied"; "Southern" researchers appeared every year, they said, only to explore an issue others had already studied in detail, and most of them left once they got the data they were after, never to return to the community. Among the critical questions from the elders and the council members were where the samples would be analyzed, what would happen to them once they had been processed, how personal privacy would be secured, how findings would be communicated to the local communities, and what financial motives the researchers might have.

To advance my understanding of local context, I interviewed a few Inuit elders with the aid of a local interpreter and requested collaboration with the Kitikmeot Heritage Society (KHS). The elders expressed interest in exploring relations among Inuit and between Inuit and other people, Tuniit and Qallunaat (whites). The KHS agreed to collaborate and to arrange for sampling. Once Helgason and I were granted the necessary permissions and the local contacts had been established, our collaborators in Cambridge Bay and Greenland collected mouth swabs, which were then shipped to Iceland. In Greenland, the procedures were different. Originally, the Greenlandic Home Rule Government wanted to take responsibility for research negotiations and licensing, but, faced with a growing bureaucratic burden, it relegated licensing to the Danish Polar Research Centre in Copenhagen, which had previously dealt with such tasks.

As I was not present during sampling, I cannot provide detailed information on its context. My impression, however, is that in Cambridge Bay people were generally enthusiastic about collaborating with the KHS. The KHS would approach the individuals in question, focusing on elderly people whose parents and grandparents had grown up in the region of the Copper Inuit. In Greenland, a physician based in the capital town of Nuuk who was involved in medical research throughout Greenland carried out sampling. The key issue was "health" rather than "heritage," but here, too, the study seems to have been generally welcomed.

A year after the sampling, Helgason and I visited Cambridge Bay and Nuuk to communicate our tentative findings and to discuss their implications with students, teachers, community leaders, and the local media. In both places, eager discussion of the results reflected the great local interest in exploring the history of the communities and of the Inuit in general. Some of our conclusions, especially those indicating pan-Inuit genetic connections throughout the Arctic and the virtual absence of early European traces in the genome of the inhabitants of the Kitikmeot region, were met with enthusiasm. When the elders of Cambridge Bay were presented with these findings, some of them cheerfully raised their fists into the air with a loud "Yes!" as if their favorite athletic team had won an important victory.

During the return visit to Cambridge Bay, we were told that one of the local concerns informally expressed when I was negotiating for research permission was potential "whiteness." Had there been a substantial "white" component in the genome, some people hinted, the community might lose its entitlements under Nunavut rules. Important indigenous rights, including health benefits and social services, might be jeopardized by genetic analyses. However, while some indigenous groups may restrict membership and entitlements (for example, revenues from casinos) with reference to genetic results, the chance of being denied Inuit status in Nunavut is negligible because, according to the Land Claims Agreement, beneficiary rights are based on self-identification: To be Inuit, it is essentially sufficient that a person be known to have an Inuk kinsperson. Moreover, to most people in Cambridge Bay, the probability of discovering hidden traces of "whiteness" in their genome seemed remote, if not an absurdity. What if our results had violated their sensibilities? The issue of "whiteness" indicates the importance of "thick" description and forms of collaboration.

The procedures followed by the IGHP are routine for research in Nunavut, Yukon, the Northwest Territories, and indeed much of the Inuit region. In hindsight, however, collaboration with the communities in question right from the beginning and more intensive fieldwork would have been desirable. While I had done research in the Northwest Territories, I had no prior experience of either Nunavut or Greenland. In her review of genetic studies of indigenous populations in northeastern Siberia, Rockhill (1999: 77) calls for an increased awareness of the implications of ethnic identity for procedures of sampling and for a new method of research "using the 'participant observation' method, [to] study the social background, life and family histories, and concept of ethnic identity, thus collecting information that emerges slowly and is not disclosed readily to a stranger." One responsible and informative way to proceed is to engage in participatory projects with "nonprofessionals." Both ethnographers and archaeologists have increasingly adopted such an approach in Arctic research.

4.4 Research practices: the Arctic and beyond

Collaborative approaches are not new to anthropology; for decades there has been much talk and experimentation on rapport, dialogue, and consultancy, partly drawing on feminist scholarship and critical ethnography. Heath (1997: 68), for instance, suggested the notion of "modest interventions" to refer to "translocal engagements that reveal, perturb, and perhaps transform the constructed boundaries between local, situated knowledges." However, "collaboration" is being formulated with more force and precision now than it was before. Lassiter (2005) argues that ethnographers are increasingly applying the term to the entire process from research design to the dissemination of results. For him, this has to do with the shifting context of fieldwork and the growth of multi-sited ethnography: "In sum, the "new ethnography" potentially moves collaboration from the taken-for-granted background of ethnography to its foreground" (Lassiter 2005: 72).

Taking the collaborative model seriously may necessitate innovative thinking and radical restructuring of academic institutions and research practices. Clifford (2004) has explored collaborations between academics and Alaskan Natives on heritage projects, pointing out that some lean toward a multivocal model juxtaposing Native and non-Native views while others are more community-based, primarily reflecting Native perspectives. In his discussion of the work of anthropologist Ann Fienup-Riordan, who collaborates closely with the Yup'ik people, Clifford (2004: 12) comments, "It is arguable that her choice to remain unaffiliated with any university or governmental institution has given her the flexibility to pioneer collaborative styles of work, engaging in relations and projects which might have seemed 'unprofessional' before they became, under pressure, the norm." Several ethnographers with firm institutional affiliations have, however, developed equally innovative collaborative styles in Arctic research.

In principle, similar arguments should apply to comparative studies of the human genome and genomic anthropology. Despite all the debates on the HGDP and the lessons they should have provided, a meaningful discussion of the key issues involved – in particular, forms of collaboration – seems strangely absent in recent reflections on the state of the discipline.

While Stone, Lurquin, and Cavalli-Sforza (2007: 233) usefully outline the issues that some critical voices have referred to as "identity politics," they conclude that a radical distinction between "scientists" and "people themselves" must be maintained: "How individuals or groups choose to use or not use genetic knowledge to construct their own kinship, ethnic or any other social or cultural identity is independent of genetic science." In light of the history of the "race" concept and its messy co-construction by scientists and the public despite the appearance of scientific "purity," such a distinction does not make much sense. As Palmié (2007: 207) has argued with reference to the use of DNA testing in "discovering" genomic pasts (in "racecraft," the making of racial histories), "Genomics, like divination, gives material shape to, and thereby reproduces as social reality ... ideologies of invisible essences and agencies."

Collaboration is not a smooth or easy road. Difficulties may stem from disagreements about fundamental premises. Some Native Americans, for instance, suggest that established archaeological narratives about the crossing of the Bering Strait are seriously flawed. Adherence to those narratives may cause havoc in collaborations on genetic history and human migrations. Another potential problem relates to the linking of rights and "race." In New Zealand, recent government settlements for grievances relating to the history of colonization have made the establishment of ancestry a central concern for the indigenous Maori. Keeping in mind increasingly strained relations between museums and indigenous groups over ancestral property, anthropologists may avoid taking any part in the forensic work involved. At the same time, given the notions of the Maori about genealogical continuity and spiritual links to the ancestors and their interests in determining whether physical remains are indigenous, some anthropologists have been led to believe that they are obliged to offer their services. Cox, Tayles, and Buckley (2006: 869) suggest that the "identification of the race and therefore the ancestry of human physical remains by forensic anthropologists could be seen as ethical responsibility." This, they argue, is not as straightforward as it may sound; the "racial" identification of individuals necessitates a set of classificatory standards that can be established only through a prolonged study of Maori remains, which is distressing to many Maori.

A further problem may be the technical nature and novelty of genomic research. It may be more difficult to elicit statements from Inuit elders about haplotype clusters and gene frequencies than, for instance, about house remains, linguistic utterances, human migrations, or soul-names. Not only approach and epistemology but also interests and motives may differ between the Inuit and non-Inuit scientists. As a result, research may mean very different things to these two groups.

Some collaborative genome projects may be abandoned at the design stage as a result of a lack of mutual understanding of goals and procedures.

Researchers may be locked in biopolitical battles that originated elsewhere, making collaborative research practically impossible. Other projects may have trouble with communication breakdowns in midcourse or at the writing stage. With the growth of indigenous movements, increasing demands for the return of ancestral bones and the repatriation of artefacts, resistance to domination and exploitation, and the quest for freedom to construct heritage and cultural identity, studies of ancestry (genomic and otherwise) are likely to be more responsive and collaborative than those in the past, exploring forms of organizing and governance suitable for different themes and contexts. One useful source of models is recent theorizing on scientific and technological development emphasizing "mode-2" forms of knowledge production (Nowotny, Scott, and Gibbon 2001) that involve heterogeneous teams geared to specific contexts and issues, in contrast to the "mode-1" forms of investigator-initiated and discipline-based approaches.

Among the questions that must be addressed in collaborative research are the following: How is a community defined, and how can researchers deal with differences in perspectives and interests? To what extent should the agents of collaboration be the pan-Inuit "community" of the Arctic, the democratic bodies of Nunavut and Greenland, individual towns and villages, social groups within them, or single individuals? What kinds of decisions should be made at different levels? How should decisions be made regarding licensing and the distribution of income generated from collaboration? Finally, keeping in mind how sensitive the issue of genetic origins and identities is, how should projects deal with the potential clash, in both research design and interpretation of results, between the perspectives of researchers and the ethnopolitics of the people providing genetic material and information?

Forms of collaboration will depend on context, researchers, and the kind of research in question. For instance, projects may differ in that some try to document human genome diversity while others seek to trace origins and migrations of particular clusters of genetic traits. The global genome, however, is riddled with major divisions and inequalities. Thus, the notion of "health biotechnology" is usually raised in association with research-intensive Euro-American universities, not in connection with poor people in developing countries. Much genomic research has assumed a rigid dividing line between the Global South (and the High North) and the Global North or the "West." Despite the recent blurring of the directions on the compass represented by these terms, the dividing line has remained rather simple: "We" deserve personal autonomy while the "natives" need to be protected by specific research agreements and instruments. In the Global North, the principle of informed consent has invariably been applied, while genome research among marginal groups in the Global South, for instance, in the work of the HGDP, introduced the concept of "group consent" (see Reardon 2005). Both principles have their rationales. The former was derived from the Nuremberg Code of 1947 and its focus on voluntary consent in biological research on human subjects, while the latter was an attempt to respond to accusations leveled at human genetic research in the Global South of failure to respect the will and dignity of the subjects. It is difficult to see why there should be such an enormous difference between Global North and South. Group consent, besides being problematic in that it naturalizes "ethnic" units and boundaries, also seems to paternalistically undermine individual rights that are taken for granted in genomic and medical research in most Euro-American contexts. Individual consent, however, has also come under fire in recent bioethics debates, indicating a communitarian turn. Knoppers and Chadwick (2005) suggest that a new "participatory approach" has emerged as a result of the growing influence of social science on ethics and the reinterpretation of the concept of "expertise" in genetic ethics.

It is important to rethink Orientalist structures of research, with their dualisms and hierarchies. It may be essential to relax or destabilize some of the ethnocentric assumptions of the grand narratives of Western biology and bioethics. Deliberately juxtaposing particular aspects of her ethnography from a Papua New Guinean context and headline news in Europe and North America about biomedical innovations and public responses to them, Bamford (2007: 176) concludes that the issues that have "recently confronted European and North American audiences concerning the dislodging of long-standing conceptual frameworks" have "little in common with Kamea perceptions of the world." For Bamford (2007: 79), Kamea notions of life itself offer a theoretical glimpse into the biosocial realities of the contemporary world: "In a world where embryos can be 'put on ice' and the dead can be forced to procreate, Kamea furnish us with a new perspective upon which to reflect upon what is essential about persons, reproduction, and crossgenerational relationships." Inuit contexts, as we will see in the following chapter, are equally informative.

The key problems encountered by the HGDP are unlikely to wither away or drop out of public consciousness unless comparative anthropological research on the human genome undergoes a paradigm shift. The ruptures and inequalities of the past should not blind us to the possibility of dialogue and collaboration in research, be it in history, archaeology, linguistics, or genomic anthropology. An important avenue for avoiding tension over representation or potential exploitation and abuse is to adopt a collaborative research model, in which subjects become consultants or even co-researchers.

4.5 Conclusions

Although the relations between the local and the global in genomic research are often fraught with tension, competition, and mistrust, there is also plenty of scope for mutual collaboration. Thus, analytical work carried out by individual laboratories benefits from the mapping of the human genome projects, and likewise, individual laboratories sometimes contribute their own mapping efforts. Such exercises underline the interdependence of individual genome projects, the new networks of associations that have been developing, and the relevance of anthropology. In some ways, the IGHP, co-organized by an anthropologist (Helgason) engaged in a variety of projects on genomics and ancestry at deCODE genetics, is an indirect product of these networks.

In many contexts, genetic discourse has a powerful role to play in informing public consciousness, reshaping existing academic fields and disciplines, and signaling the arrival of new ones. The discourse, however, is complex, and a new paradigm may be imminent. As Lock (2005: S51) suggests, "It is eminently conceivable that a paradigm shift of enormous significance is now under way in basic biology, a shift that could potentially transcend outmoded nature/nurture debates and simplistic discussions of gene-environment interactions." Anthropology cannot resist being affected by and affecting these developments. Not only does the new genetics provide an exciting site for the ethnographic exploration of a whole range of issues, many of which are identified with reference to the Greek term for "life" (biopower, biosociality, biopolitics, biomedicine, biovalue, and so on), but it also opens new and important avenues for studying human variation, history, and evolution.

I have suggested that anthropology must seriously engage with genomic research, resisting both the naturalization of the social and the socialization of the natural. In their attempt to "unwrap" the "sacred bundle" of the four-fields approach in anthropology, Segal and Yanagisako (2005: 11) ask the following rhetorical question of colleagues who argue for holism: "When was the last time that research on hominid evolution or primates was helpful to you in thinking about your ethnographic data?" The point is not to reinforce a rigid monologue along the lines of the traditional four-fields approach but to foster an open-ended, hybrid network of subdisciplines, programs, and research practices suitable for an academe that is increasingly postdisciplinary. Such an approach seems to resonate with a "world anthropology" (Restrepo and Escobar 2005) that attempts to break out of the mould of European and North American contexts.

Introduced by Hans Winkler in 1920, the word "genome" is a portmanteau of "gene" and "chromosome." While the genome concept itself draws attention to neither epigenetic processes beyond the cell or organism nor critical engagements with the larger context of genetic research, it does seem to carry an implicit reference to some kind of structure. Moreover, it resists the reductionism implicit in the alternative labels of "biological" and "genetic," allowing for causality in development either within or outside the genome or at the juncture of the inside and the outside. In sum, genomic anthropology involves the study of human genomic variation and history; it attends to basic concepts and metaphors, indigenous and otherwise, including population, lineage, race, kinship, and soul-name, and engages with the implications of genomics for biopolitics and the construction of relations and identities. Genomic anthropology is not a prefabricated thing finally revealing itself to the larger world but rather an important avenue for resolving some of the tensions associated with the study of humans, partly through serious engagement and collaboration with the subjects being studied.

It is too early to say on what terms anthropology will relate to the new genetics. I suggest, however, that anthropologists are co-constituting fresh networks among scholars, laboratories, private agencies, and national and multinational agencies (see Heath 1997). One network is the multinational GenBank, an online genetic sequence database comprising all publicly available DNA sequences. Another important development has been the International Human Genome Project. Despite important ethnopolitical stumbling blocks, the project has generated new bodies of anthropological data. Moreover, it has increasingly brought anthropologists into the realm of biomedical research, where their understanding of human differences and history is useful in population-based studies, the interpretation of comparative data, and the construction and use of large-scale genetic databases.

Perhaps academic fields and (sub)disciplines are best seen as imagined communities. Their practitioners do not necessarily meet face-to-face as a collectivity, nor do their trades consist of predetermined essences destined to unfold through the processes of history. Their communities are constructed and maintained through the boundary work of forums and agencies: conferences, journals, websites, departments, and laboratories. The term "genomic anthropology" will gain currency only to the extent that the anthropologists involved in genomic research find it useful and imagine themselves as belonging to the new networks of associations within which their work takes place.

5 Name talk: technologies of belonging

In the social world constructed in the play *The Numbered* (1984) by Elias Canetti, the citizens are allocated names according to the number of years they will live ("Fifty," "Twenty-Five," "Twelve," and so on). No one, it is assumed, is able to escape his or her specified fate. The names are decisive and fatal and the subjects are not consulted. One of the persons in the play is worried about "Twelve," who has run away from home:

> She was afraid of her name. People told her she had to die when she was twelve . . . She grew quieter and quieter. We didn't know why she spoke so little . . . But then, on her birthday, fear seized her. She disappeared. She went away among people who did not know her name . . . Since then she has been hiding . . . She has avoided us like the plague. (Canetti 1984: 94)

It seems likely that Canetti developed his name talk as an analogy for the gene talk on the "secret of life" gaining currency at the time of writing, about a decade after the discovery of the double helix. Perhaps one should see the play as a commentary on the idea that our lives are more or less determined by both our names and our genetic characteristics, a theme popularly explored, for instance, in the film *Gattaca*, which radically separates "valids" from "in-valids." Canetti's names are like ticking time bombs: 12, 11, 10, . . . Bang! On their own, names have significant impact on our lives and, as a result, practices of naming represent a particular form of biopolitics, a technology of the self. Sometimes names imply deliberate exclusion or social death, commencing at the time of naming when the name begins to operate in the manner anticipated.

Because they are essential to every person everywhere, personal names have often been taken for granted. Anthropology, sociology, and philosophy, however, have created a fairly extensive literature on the variety of systems of naming in different times and cultural contexts (for a useful recent overview, see vom Bruck and Bodenhorn 2006). Drawing upon this literature one may ask: To what extent do naming practices exemplify or parallel the biopolitics of bodily inscriptions and markings such as tattoos and presumed racial signatures? Paraphrasing Foucault, it seems that names not only specify and individualize their bearers, they also represent "technologies of the self"

(Foucault 1988), serving as means of domination and empowerment, facilitating collective action, surveillance, and subjugation – exclusion as well as belonging.

Practices of naming have probably followed humans since the development of rudimentary language. Yet, they vary considerably from case to case and from one time to another. This poses a number of questions: Why, in particular, would humans bother to adopt names? What kinds of choices are available in each case? What kinds of roles do names play in social life? Why do traditions of naming vary from one context to another and from one epoch to another? What are the conditions and processes through which names acquire their momentum or "illocutionary force," in Austin's sense (engaging in "performance," "securing uptake," and producing "conventional consequences") – guaranteeing what Pina-Cabral (2011) refers to as the "ontological weight" of names?

Personal names, as Lévy-Bruhl observed, are not simply neutral devices for classification and indexing but are inseparable from the person, embodied and emotionally charged; people tend to "regard their names as ... sacred... For such a person, a name is a distinct part of his personality, just as much as are his eyes ... and [he] believes that injury will result as surely from the malicious handling of his names as from a wound inflicted on any part of his physical organism" (quoted in Bodenhorn and vom Bruck 2006: 9). Drawing on Lévy-Bruhl's perspective, Bodenhorn and vom Bruck (2006: 20) suggest that it is "particularly striking that names have not featured in analyses of the body because they engage all crucial aspects of embodied experience: identification, moral relations, power, the gendering and sexualizing of bodies, and displacement."

This chapter discusses personal names in relation to other technologies of belonging, highlighting the relevance of practices of naming for body politics, radical differences between the recent Western tradition of surnames and some other naming traditions and epistemologies, the formation of subjectivity in slavery through renaming, and the cultural and political clashes that sometimes result when an expanding empire encounters and colonizes an indigenous tradition. Practices of naming, it is argued, are not only key elements of identification and personhood, embodied in the biosocial habitus much like other biomarkers, they also situate people in genealogies, social networks, states, and empires. Naming involves powerful speech acts, making history, constituting persons and the social relations and systems within which they are embedded – families, communities, states, and empires; indeed, naming is one of the key examples used in Austin's (1962) classic formulation of speech acts.

While for many Inuit the role of personal naming is similar to that of genes in modern gene talk in the sense that the individual is seen as a vehicle for hereditary information, there are important differences. Names, unlike the genes of gene talk, are transferred horizontally as well as vertically through human agency, the speech acts of relatives, friends, and neighbors – and the person adopting the name. Such a comparison, besides being worthwhile as an analytical method, may also help in the understanding of Inuit comments on genomic studies and the potential impact of genomic studies on indigenous notions of sociality and citizenship. Clashes, I suggest, between different traditions and practices of naming, especially in the context of slavery and empire, illuminate with striking clarity the relevance of names as technologies of exclusion and belonging. Further explorations should elaborate on the differences as well as the parallels of name talk and gene talk and the multiple forms and social implications of both kinds of discourses.

5.1 Technologies of naming

Despite differences in tradition, certain fairly obvious generalizations can be made about personal names. They always serve the purpose of situating people in social space, connecting them to family, lineage, ethnic group, and so on. The practice of naming, however, is not simply a classificatory exercise. Naming is a speech act, shaping the life course and the person involved. Names, in other words, both personify and embody, resulting in durable dispositions in Bourdieu's (1978) terminology. The reason why names "stick," why the speech acts work, is that somehow the acting speaker is granted the license to name by the community involved, through a formal or informal social contract. That "somehow" is rather difficult to establish or define. The social contract is sometimes based on relations of relative equity and sometimes on relations of subjugation and power (typically for derogatory nicknames). Either way, without the contract the person would not adjust to the name and, as a result, the name would simply be discarded like worn or irrelevant clothes.

Any attempt to account for the force of personal names has to attend to the notion of context. Goodwin and Duranti (1992: 4) point out that a relationship of mutuality is "central to the notion of context (indeed the term comes from the Latin *contextus*, which means 'a joining together')." Context and talk, they suggest, "stand in a mutually reflexive relationship to each other, . . . talk, and the interpretive work it generates, shaping context as much as context shapes talk" (Goodwin and Duranti 1992: 31). Just as namers and the names they establish are inseparable from the community in which they are embedded, the person is inseparable from his or her environment. Subjectivity, identity, and naming, then, are informed by the social and political environment in which the person is embedded (Jackson 2005). This is why naming practices are often a contested issue for groups campaigning for human rights and social justice; renaming involves redefining context.

Patrilineal surnames, an influential technology in the modern world, developed in Europe during the Middle Ages. Along with the genealogical tree, they served the purpose of tracking time, succession, and inheritance among agrarian elites, documenting at the same time the symbolic and social capital on which political alliances were based. Pina-Cabral (2011) explores the discourse and reality of "true" names in Portuguese-speaking contexts, partly in the context of a shift in the allocation of administrative responsibilities early in the twentieth century as the national Civil Registry took over some of the roles traditionally played by the ecclesiastical register. "In the Lusophone tradition," he argues, "'truer' names rise above the flux of everyday relations . . ., they give an appearance of permanence and inalterability that makes them indispensible instruments of bureaucratic control" (Pina-Cabral 2011: 308). In other words, the "true" names sanctioned by the overarching authority of the state and its Registry determine the subjectivity of national citizens, operating as a technology of belonging.

Patrilineal surnames are not universal in the Euro-American context. Thus, for instance, Icelanders generally do not have family names, in line with an old Norse tradition. People are listed and known, above all, by their first names, which embody their persona. The second name is usually a patronym (matronyms, however, are increasingly used, either on their own or along with the patronym). "Jónsson" (or "Jónsdóttir"), for example, simply indicates that the person in question is the son (or daughter) of a man named "Jón." Women rarely change their name when married, nor do men. This, however, has not created major administrative hurdles, nor has it prevented or slowed down the development of population projects and genomic studies (Palsson 2007).

While patronyms and matronyms do not provide the same level of cultural and genealogical memory as patrilineal surnames, they nevertheless ensure some attention to the history of social networks. A striking contrast to the excessive cultural memory generated by patrilineal surnames is provided by the teknonymy of Bali. This system consistently erases knowledge of collateral kinship ties with each generation, creating "an ever-repeating sequence of four impersonal statuses – child, parent, grandparent, and great-grandparent" (Geertz and Geertz 1964: 102). Such genealogical amnesia is not just an odd ethnographic curiosity, it represents "a 'cultural paradigm' of social roles and relationships which serves the Balinese as an interpretive guide for understanding and manoeuvring within their own institutional system and as a set of instructions, a programme in the computer-theory sense, for maintaining that system" (Geertz and Geertz 1964: 103).

One of the Orientalist structures of modernity is the hierarchy of expert and lay knowledge. Given that the separation of the natural and the social is ethnographically and theoretically suspect, such a hierarchy is not helpful in research. In a collaborative approach, the real challenge for anthropology is to explore and engage with the theorizing of the people being studied. Such engagement is common in anthropological studies of science, which "study up," but it has been less common in other contexts. While studies of genetic ancestry tend to generate convincing narratives, the "biological facts" they construct in probabilistic terms through laboratory work need not have privileged epistemological status (Latour and Woolgar 1979). It is important to recognize that our informants may see things differently. To explore patterns of regional movement and differentiation, the organizers of the Inuit Genetic History Project discussed in the previous chapter asked those who provided genetic samples to fill out a brief questionnaire specifying, in particular, the birthplace of their grandparents. Since, for many Inuit, people are rooted in the land, it is unlikely that the respondents had significant doubts about the places in question (significantly, the suffix *miut* [people], which identifies families and communities with the places they occupy, has become co-opted in current ethnopolitical discourse). However, respondents in Kitikmeot and Greenland need not have given priority to what would normally be described as "biological" grandparents. What, then, do Inuit make of inheritance, relatedness, and sociality, and how does their theorizing relate to that of others?

5.2 Inuit name talk

A glimpse into the ethnography of indigenous communities in North America helps to further establish a radical contrast with the system of patrilineal surnames, with which Euro-American readers may be most familiar, and to set up a useful context for exploring culture clashes heavily focused on renaming. Here as elsewhere, practices of naming are firmly rooted in epistemologies of belonging, relatedness, and becoming human.

It is difficult to generalize about notions of sociality in the Inuit region; different ethnographers do not always agree, and, moreover, the cultural and geographical terrain from Alaska to Greenland is vast and diverse. The ethnographic present adds further complications, if only because the arrival of Christianity and empire has had a profound but variable impact. Inuit kinship appears to be changeable and flexible, like other social connections. Adoption is fairly common, and often children move, temporarily at least, from one household to another. Inuit do differentiate between adoptive and "biological" relationships (see Burch 2006: chap. 3); when, for example, attached to a root kin term such as *ataata*, the suffix *saq* indicates an adoptive relationship: *ataatasaq* (adopted father) or *ilniqsaq* (adopted son). Alia's (2006: 36) remark that "in the Inuit custom-adoption culture the whole Qallunaaq [white] idea of biological versus adoptive families has no relevance" is therefore an overstatement. At the same time, a man who has genetically fathered a child is its "real" father only to the extent that he establishes paternity by his actions, by looking after the child and caring for it after its entry into the world.

The complex social network that forms an Inuk's person is to a considerable extent interlinked through personal names. Persons are called into being when a soul-name (*atiq*) is spoken. While the idea of the soul-name can be interpreted in several ways and it is not clear what is transferred, naming seems to create multiple personalities, particularly through the combination of names: "You are not simply playing different persons; you are different persons" (Bodenhorn 2006: 151). For the Iñupiat of Alaska, names are "open" in the sense that while they do have a social life and travel through time, they usually do not classify nameholders in terms of gender, kinship, or social position. Nuttall (1992) gives a similar interpretation in the Greenlandic context, suggesting that this naming practice pertains to most Inuit. Names imply certain personality traits that are passed from one person to another, traits that are recycled with each new generation. Relatives, friends, and acquaintances give each other names both as children and later in life.

Several ethnographers have underlined the central role of naming in the Inuit context. In Alia's (2006: 37) summary, "naming is a – perhaps *the* – central component of Inuit culture." Many observers have nevertheless been baffled by the complexity and apparent contradictions of Inuit naming practices. Stefansson's field notes from a century ago provide one illustration. He had often heard a particular woman address her daughter Noashak as her mother. Then, he wrote, "Today we had a curious example of the final grasping of an idea which one would think should have come years ago" (see Palsson 2001: 285–286). It had finally dawned on him that this naming might be seen as a form of reincarnation theory, Noashak being understood as two dead individuals.

It is illuminating to explore the similarities and differences between Inuit name talk and modern gene talk. In some ways, the role attributed by many Inuit to personal names is similar to that of genes in the program theory. Names, like genes, somehow inform or establish identity and personhood, accounting at the same time for divergence and continuity. Indeed, when describing Inuit realities, some Inuit and ethnographers have resorted to a quasi-genetic language of "vehicles," "mutations," and "substance" similar to that of mainstream genetics. Thus, as one of Bodenhorn's Alaskan Inuit informants suggested, a child is "a vehicle for the name" (Bodenhorn 2006: 147). Similarly, Williams argued that the Inuit "soul complex and naming system is in fact a series of constantly-changing transcendencies and mutations" (quoted in Alia 2006: 22), and Saladin d'Anglure (1977: 33) reported the Inuit belief that "the essential ingredient of a human being is its name. The name embodies a mystical substance which includes the personality, specific skills, and basic character which the individual will exhibit in life."

Inuit sometimes refer to personal names when explaining behavioral traits that tend to be medicalized in contexts heavily informed by gene talk. Thus, hyperactivity ("attention deficit hyperactivity disorder" in medical language) in children is sometimes accounted for by their particular names or the fact that they have too many names or a name that does not "fit." Stefansson observed that restless Inuit children sometimes seemed to protest against their names, requesting renaming (see Palsson 2001: 156).

While for many Inuit the role of personal naming is similar to that of hereditary material in modern gene talk, there are important differences. Essentially, Inuit discourse on identity and relatedness is nonreductionist and relational, in line with the principles of epigenetics and developmental-systems theory. Inuit names operate much like horizontally transferred genes, cosmopolitan names passing from one person to another depending on a host of personal and contextual factors. Also, and this follows from the notion of agency, names may change during the life course, remaking personhood in the process. This does not mean that Inuit kinship is simply a matter of "cultural" construction or framing. For Inuit and other epigenetic theorists, "biology" (in the conventional Western sense) is beside the point; fatherhood and motherhood are always "real" and embodied. In Ingold's (2001: 257) approach to kinship, which aims to dissolve the boundary between "biology" and "culture," behavioral dispositions are "formed in and through a process of ontogenetic development within a specific environmental context. Kinship is about the ways in which others in the environment contribute – through their presence, their activities and the nurturance they provide – to this process." This resonates with Inuit thought. It is partly through naming that Inuit children are positioned in a relational field through which their biosociality unfolds.

Not only is the person constituted by names, names are the subject of constant discussions as, quite literally, technologies of belonging. Among the Yup'ik of Alaska, children are usually named after a recently deceased relative or community member. The bestowal of a name is marked with offerings and gifts and, in the process, the person becomes more than a relative. The ceremony during which a child is named is called *kangiliriyaraq* which literally means "to provide with a beginning" (Fienup-Riordan 2000: 192). For the Yup'ik, personhood would not be generated without parents and biological birth, but what matters above all is ancestral names defining a person's position within a particular genealogy. Aside from ancestral names, people also have "calling names," many of which are teasing names similar to what is usually referred to as "nicknames." A further ethnographic case is that of the Tsimshian of northwestern British Columbia. For Tsimshian, the act of naming "gives the person to the name" (Roth 2008: 15). Proper names are selected from a "basket" of "vacant" names. Reincarnation is of central importance, "an undeniable fact of the universe ... It is a fact of nature rather than an aspect of their 'culture'" (Roth 2008: 62). The English term "reincarnation" is in fact a translation of indigenous terms denoting salmon "running together" to or from their spawning grounds year after year.

Yup'ik, Tsimshian, and Inuit notions of sociality and personhood, evident in much of their name talk, highlight the irrelevance of the idea of the autonomy of the "biological" as commonly understood. The biology of the organism seems inseparable from the durable dispositions of the habitus developed in the course of everyday practices. Although indigenous North American notions of naming and personhood tend to be anathema to genetic concepts of heredity, essence, and relatedness, they have a clear bodily reference. Names embody what needs to be embodied, fashioning the organism and the person, which, in the local view, is the same.

5.3 Renaming

While modern states and empires encourage and sometimes enforce stability of names, assuming the same name from birth to death, names frequently change. In many contexts, people become known by nicknames that sometimes overshadow official names. Nicknames can be purely classificatory, distinguishing between people with almost identical formal names, but often they locate people in social space, much like other names. Occasionally, a nickname begins to operate as a "surname," applied to a whole family and extending across generations. Sometimes nicknames are denigrating, the result of some kind of competition or power game, moulding the life and persona of the individuals to which they are attached. Like other names, they depend on a social contract without which they would have no force or weight. In extreme cases, the persona is severely injured as a result of a nickname. The wounds inflicted in the process may not be highly visible, although sometimes people clearly "lose face," and can be just as serious as physical violence. Indeed, this *is* physical violence in a very real sense, with lasting and damaging embodied results. The *abuse* of names, harassment in the form of nicknames, testifies to the bodily reference generating sensation and shame.

Some of the most extreme cases of renaming come from the history of slavery. Slaveholders are usually keen to rename their slaves, often with names not unlike those applied to pets and livestock. At the same time, the application or citing of the person's original name is liable to be subject to punishment. Thus, the persona of the slave is deformed with the new name, torn from its former social environment: Not only is the persona of the slave eradicated, the slave is "marked off from other persons whose social identities are given privileged recognition" (Benson 2006: 181). Significantly, when slaves are granted freedom they often insist upon formally receiving a new name in front of witnesses, to mark the ending of oppression, to regain dignity, and to publically confirm the support and acceptance of society.

Caribbean slavery is a case in point (Palsson 2016). Most of the slaves in the Virgin Islands came from West Africa. The slave owners usually adopted new kinds of names for their slaves that would have sounded alien to their free West African ancestors. Slaves had to accept being renamed. This was part of the erasure of identity and history that characterized the so-called "Middle Passage," between West Africa and the plantations (Davis 2006). While slave owners had complex and colorful names indicating social connections across time and space – one example being "Heinrich Ludvig Ernst von Schimmelmann," the name of a Danish aristocrat and plantation owner, a key player in the plantation economy of Virgin Islands – their slaves were only called by one or more "first" names that distinguished them from each other, and all of them, by the absence of a "second" name or a family name, from their masters and other free persons. The slave names on eighteenth-century plantation lists from the island of St. Croix are usually European or Western: Emilia, Regina, Hans, Jonatan, Anna, Maria, Andreas, and so on. Occasionally, though, one can see "strange" names on the tax lists, exotic to Euro-American readers: Profix, Polepti, Leipis, Suatre. Probably these were recent "acquisitions" that had not yet been renamed.

Sometimes Caribbean slaves were given a name associated with people of historical importance, simply to humiliate them and keep them under control. The name of Cicero was one of the popular names. A slave carrying such a name was constantly reminded of his marginal subjectivity. One of the house slaves of plantation-owner Schimmelmann mentioned above first appears in the slave records as "Cicero." Later, however, at a baptism ceremony in St. Croix, he was renamed "Carl Heinrich," probably testifying to a significant change in status.

The cases of the Yup'ik and Inuit illustrate what may happen when a powerful and highly orchestrated system of surnames, sanctioned by an expanding state, clashes with a radically different set of naming practices, in the absence of surnames. In the Yup'ik context, particular administrative and legal complications in documenting identity have sometimes been associated with multiple names. How would one issue a passport when a person's records – baptismal and birth certificates, government enrolment forms, and so on – have been issued under a series of different names? This was a question the US Passport Agency was bound to raise. As late as 1997, it initially denied requests for passports for four Yup'ik elders who planned to travel abroad and finally issued them only on a limited basis (Fienup-Riordan 2000: 195).

The customary naming practices of the Inuit often clashed with bureaucratic procedures in Alaska, Canada, and Greenland that sought to "tidy up" Inuit nomenclature and normalize a standard patriarchal family system deemed well suited to national citizenship and civilized conduct. In Canada, a system of identification by means of serial numbers was proposed in 1935 to craft legible identities for the Inuit population and to avoid the "confusion" of native practices, which baffled the ruling officials. A medical officer originally proposed the number system with the "humble suggestion" that "at each registration the child be given an identity disk on the same lines as the army identity disk and the same insistence that it be worn at all times. The novelty of it would appeal to the natives" (see Alia 2006: 52). Inspired by the military use of dog tags, this system suggested that a number be engraved on the disk to facilitate easy identification by outsiders. This would make it possible to monitor the Inuit as national subjects. Scott, Tehranian, and Mathias (2002: 27) speculate that had the technology of the time permitted, "there is little doubt that the officials would have preferred small electronic transmitters and global positioning systems to monitor all movement by satellite." Undoubtedly they would have preferred genetic signatures and DNA fingerprints as well. Enforced renaming along the lines of the disks (*ujamiit*) is a violent act that necessarily downgrades the people involved, a frequent outcome of slavery and empire.

Partly as a result of complaints of disrespect and internal colonialism, the disk system was abandoned in 1968. During the past century, a system of family names was introduced throughout most of the Inuit regions of Alaska, Canada, and Greenland, at different times and paces in different places, partly as a result of the growing number of children of mixed Inuit-white parentage. In Canada, there was a contrast between west and east. In the east (roughly the future Nunavut), Project Surname was introduced in 1969, alongside the customary naming system, partly to replace the disk system. Currently, name pluralism is typical for Inuit throughout the North, with the systems of patriarchal surnames and customary names coexisting in many places, their use dependent on context. There is much interest in keeping customary names as part of "tradition" in Nunavut, and they are still quite common there.

Given the importance of naming in Inuit contexts, the imposition of the disk system and patriarchal surnames can be regarded as a biosocial undertaking analogous to genetic engineering. Such changes helped to reconstitute the Inuit subject in the dual Foucauldian sense: governing the people (minimizing or eradicating what Canadian officials tended to see as a hopeless administrative mess) and redefining their biosociality (refashioning their habitus – that is, their relatively durable embodied dispositions). The refashioning of subjectivities, bodies, and groupings through acts of naming and categorization is a common predicament of people subjected to the encroachment of states. This, indeed, problematizes life, allowing for "life beside itself" (Stevenson 2014), life that requires maintenance and effort (Praet 2013).

5.4 Names, populations, and ethnic groups

Given the historical role and significance of patrilineal surnames in Europe, it need not be surprising that they still seem to provide indicators of the regional, cultural, and genetic structure of continental populations. Taking a broad geographic perspective, Mateos (2007) suggests that often people's names offer a convenient window into population structures, especially in the absence of reliable knowledge about self-identified ethnicity, and, as a result, names both open up a new era of genetic genealogy and are an important tool for policy in today's multicultural society. Patrilineal surnames, indeed, have proved a useful avenue into European genetic history, a field pioneered by L.L. Cavalli-Sforza (see, for instance, Stone and Lurquin 2005: 62).

A perennial problem, however, for social and biological analysts as well policy makers and administrators is how to define and demarcate human "populations" (see, for instance, M'charek 2005 on the Human Genome Diversity Project). While molecular studies removed anthropometry and the categorization of races to the sidelines during the second half of the twentieth century, at least in scientific discourse, focusing on gene frequencies and sequences rather than phenotypic characteristics, they tend to fall back on problematic notions of populations and ethnic groups.

Years ago, Ardener launched a critique of the bounded notion of populations and ethnic groups in demographic studies, a critique that seems pertinent to many modern studies of genomic differences and human variation: "[A]re the entities called 'populations'," he asked, "*names* or *numbers*? If names: named for whom, and by whom? If numbers: counted by whom, and for whom? In asking the questions 'by whom?' and 'for whom?' we also ask in particular: by or for the 'people' concerned? Or by or for the anthropologist or other scientific observer?" (Ardener 1989: 110; emphasis in the original). Including the human geneticist, we might add. Research on genetic history has its own politics of naming and categorization in the identification of "populations" for sampling diversity (Palsson 2007: chap. 7). One of the critical issues is potential bias in sampling and the resultant possibility of circular reasoning: Populations have been sampled on the basis of preconceived ideas about geographical distributions and differences, generating empirical results that seem to support preconceived ideas about human variation. While many reports elaborate on the laboratory techniques employed, less attention has been given to sampling procedure. The sampling in Siberian studies, for instance, based on brief and simple interviews, has been flawed; in some cases, "people's ethnic identity . . . may . . . depend on the vagaries of . . . ideological reason or . . . shifting state policies" (Rockhill 1999: 69). Frequently, a dearth of genealogical and ethnographic information has led to understatements of both the mobility and the genetic admixture of indigenous groups before and during the Soviet period.

As many anthropologists have emphasized, among them Ardener, ethnic groups are fluid units with flexible boundaries, subject to both self-identification, and naming. This is a theme classically explored by Fredrik Barth in his *Ethnic Groups and Boundaries* (1969). Ethnic boundaries are *made* for specific pragmatic purposes. Personal names, indeed, somewhat like gene frequencies, are often implicated in the process of demarcating and defining populations and ethnicities. Significantly, after the riots in Paris in 2006 which involved mostly young people of North African origin, a French security agent was quoted as saying that the rioters "feel penalized by their poverty, the colour of their skin *and their names*" (quoted in Finch 2008: 713; emphasis added). As Palmié (2007) observes, personal names are part of "racecraft" or race making.

5.5 Conclusions

Personal names play important roles in all societies at all times, although practices vary significantly both over time and from one context to another. The "strange" naming practices discussed above, the recycling of personal names among Inuit, help to de-familiarize the contemporary systems with which most Euro-American readers are likely to be familiar. In all cases, however, naming systems operate – sometimes tacitly and sometimes explicitly – as technologies of differentiation and belonging, much like racecraft and genomic essentialism. The study of names and practices of naming, as a result, should be included in studies of biopolitics. Finch (2008: 709) suggests that "sociological research on names and their use is surprisingly sparse given their social significance." While this is an overstatement, practices of naming deserve greater attention in social science and humanities research, partly in the context of the technologies of belonging.

Drawing partly on my own research in Inuit contexts, I have explored some of the opportunities offered for anthropology in the genomic domain and identified both some tensions that they invite and some potential ways of resolving them. The history of the Inuit can be advanced through crossdisciplinary perspectives, and Inuit notions of subjectivity and relatedness can also provide an interesting and useful contrast to the program theory of genes and development. For many Inuit, a person's character is determined less by the parents' DNA than by the personal names he or she is given or by the act of naming. While Inuit epigenetics tends to resist the biologizing of kinship that pervades Western discourse, for Inuit identity and relatedness are nevertheless thoroughly embodied.

Genomic studies must attend to local practices and conceptions of personhood, naming, and relatedness. Local understanding and interpretation of genetic analyses are likely to depend on a host of factors, including how people construct nationhood, citizenship, subjectivity, and relatedness. Even when researchers avoid making statements about identities and relatedness, the mere fact that they are sampling DNA in an attempt to map genetic histories may unavoidably engender genetic citizenship, possibly revising established notions of belonging. Genetic studies are, however, unlikely to replace established notions of sociality with new ones. Instead, their results will be filtered through local notions. Likewise, local notions may inform the ideas and theories entertained by visitors. There may, for instance, be a fruitful dialogue between Inuit epigeneticists and professional developmental-systems theorists.

It seems pertinent to explore the variety in current naming traditions in terms of what they do, the impact of the speech acts involved, integrating analyses of such technologies into studies of the constitution of subjects, the politics of belonging, exclusion, and control. The perspective of personal names is highly pertinent for the study of the use of signatures associated with ethnicity and race and social attempts to modify, establish, or erase any kind of social hierarchy. While names are embodied much like other biomarkers, they are not fixed, natural essences; if that were the case, people would not be renamed, and naming traditions would not change. Just as slaves can regain their freedom and personhood, those stigmatized by nicknames can regain their dignity – through revising the social contract on which naming is based. The numbered, after all, can be renumbered.

II

Bodies

In this section the focus shifts to bodies and species, moving from the fascination with border species, between nature and culture, and the cognitive abilities of non-human animals to futuristic speculations about humans in outer space. The key argument is that the issue of human essence needs to be problematized and reframed in relational, pragmatic terms. Recent theoretical developments in anthropology, sociology, and biology – developments often associated with the notions of biosociality, nature-cultures, embodiment, and entanglement – contribute to the growing evidence of an expanded and relational body. While anthropology and, indeed, most of the social sciences have, almost by definition, tended to focus on the relations of *humans*, in recent years the discursive division between humans and non-human animals has been increasingly blurred.

6 Human variation: shifting perspectives

To the extent that social and cultural anthropologists have been interested in the issue of human variation, for them it has usually referred to differences in cultural practices and social institutions – to comparisons of, for instance, "culture areas" and the mosaic, ethnographic samples covered by the Human Relations Area Files. Such comparison has been vastly complicated in recent decades by doubts about the autonomy of the samples being compared and postmodern challenges to ethnographic description. Arguably, as we will see, language, speech, and accent, and other "embodied" skills and practices studied by social and cultural anthropologists, represent aspects of human variation, no less than DNA signatures and cephalic measurements. Human variation, however, has been conventionally understood as something different, and usually left for those concerned with the "biological" domain – in both anthropology and other fields, including biology and medicine. While biological and physical anthropologists have been keenly interested in human variation, however, often treating it as the *defining* element of their subdiscipline(s), for them the scope of "difference" has tended to be taken for granted, remaining strangely inert as if the framework of observation and the parameters of variance to be explored had been fixed for all times.

What counts as difference, in fact, may not be that clear. One reason is continued uncertainty about what counts as *human*, the demarcation and definition of "our" species. Adding to these uncertainties, as we will see, a series of developments (including studies of the human microbiome and its coevolution with humans) have challenged current understandings of the boundaries and signatures of the human. What counts as "us" and what is more-than-human? Furthermore, the prospects of the *post*-human raise fundamental questions about human futures and identities. More fundamentally, I suggest, theoretical developments in anthropology, sociology, philosophy, biology, and related fields – developments often associated with the notions of biosociality, nature-cultures, embodiment, epigenetics, and entanglement – contribute to the growing evidence and perception of an expanded and relational body. Drawing upon recent works on "biosocial becomings" (Ingold and Palsson 2013), "body worlds" (Harris and Robb 2013a, 2013b), and "polygenomic organisms" (Dupré 2010, 2012), I suggest it is pertinent to rethink human variation. Not only have the notions of "biology" and "the body" been radically socialized (Meloni 2014), also the notions of "social theory" and "society" have been thoroughly embodied and materialized.

In their attempt to historicize the body and to move beyond the master narrative of the pre-cultural, biologically necessary body, Harris and Robb (2013a: 3) have developed the notion of "body worlds":

> A body world ... encompasses the totality of bodily experiences, practices and representations in a specific place and time. These ... are at the heart of how we understand the world ... for all human beings body worlds are the worlds all of us inhabit all the time. Far from being boring, natural or universal, they are in fact fascinating, diverse and culturally specific.

A social turn in physical and biological anthropology seems imminent, a new body world – a paradigmatic change in "thought style," in Fleck's sense of gestalt shift in intellectual interests.

In this chapter I discuss the histories of physical and biological anthropology, their characteristic understanding of human variation, the challenges posed by recent studies of the porous and relational body, the nature of the biosocial turn which seems to be taking place in anthropology and related fields, and its implications for representations of human variability. Given such rethinking, I argue, the exploration of human variation becomes the joint project of "social" and "natural" fields of scholarship. In such a project, however, variation should not be seen as the mapping of essences and states of being but of relations and becoming.

6.1 Physical and biological anthropology

Early on, during the age of discovery and colonialism, racial thought relegated some of us to the sub-human, often on the basis of skin color. The eighteenth-century slave trade across the Middle Passage highlighted and inscribed a fundamental distinction between free people and slaves, black and white (Davis 2006). Voltaire described "prodigious differences" between Africans and non-Africans, declaring them different "species of man" (Jablonski 2012: 136). Sub-human ideas have persisted in some circles until recent times, including Nazi Germany and Maoist China (Proctor 1988, Feuchtwang 2006).

Leaving aside uncertainties about the human, the variation *within* humanity to document and to compare continues to be contested and revised, among anthropologists, geneticists, and the general public (see, for instance, Keita *et al.* 2004 and Richardson 2013 on "race" and "sex itself," respectively). In the seventh edition of their popular textbook on physical anthropology, Stein and Rowe (2000: 433) make a fundamental distinction between folk and scientific classifications of humans, emphasizing the biases and the instability of the former:

> Scientific classifications of people, like folk taxonomies of people, are attempts to divide human beings into specific groups, but this is where the similarity ends. A scientific classification is a means of discovering the processes that create the phenomenon being classified, in this case human variation ... Folk taxonomies are usually based on ethnocentric ideas about the inherent differences in physical appearance and behavior between groups. Although some of these beliefs may be partially based on observation, most are based on folklore stereotypes. In contrast, the criteria used in scientific classification must be derived from empirical studies.

One should not forget, however, that nineteenth-century science badly mismeasured humans, misinformed by, and misinforming, the public. Observations and empirical studies cannot be easily separated from the body worlds within which they are embedded. Interestingly, one of the key illustrations in Fleck's pioneering book on thought styles, originally published in 1935, demonstrated the usefulness of his perspective through discussion of changing understandings of human anatomy. "To obtain an even clearer picture of how scientific observation differs when two different thought styles are involved," he suggested, "it is perhaps appropriate to compare anatomical descriptions and illustrations in early and recent text books" (Fleck 1979: 133). "Human variation" remains a shifting and contested terrain in both scientific and folk accounts.

For a long time the study of human bones was an important component of physical anthropological and archaeological research on human variation and history. Passionate bone collection and measurements continued in some contexts throughout the twentieth century, both constituting and being constituted by the imagined communities of nations, cultures, and disciplines and their traditions of remembrance and authority (Connerton 1989). As Verdery (1999: 27) notes, because bones are material things and "indisputably there" they have political lives of their own: "A body's materiality can be critical to its symbolic efficacy." The measurement and classification of bones became an obsession. This is underlined by the extensive cross-cultural bone collections of the Peabody Museum at Harvard University stored in hundreds of cardboard boxes in a large hall of the museum, labeled with the names of nations, peoples, and places.

Bone collections, however, would have been unthinkable, at least meaningless, in the absence of a shift in European body worlds. As Harris and Robb show (2013b), while at the beginning of the Early Modern period bodies were considered to be extremely porous to influences of environment and external events, later on, with the rise of the Enlightenment view, they were rendered as machines, with firm, policed boundaries and a specific habitat. At the same time, the idea of race acquired its full biological definition.

Skulls – a particularly important part of the body machine, as the seat of rationality, mind, personhood, and identity – were endlessly recovered from graves, measured, classified, analyzed, stored, and displayed in museums and archives.

With its set of tools for measuring human bodies, from the "anthropometron" of the seventeenth century and the "craniometer" of the nineteenth century onwards, physical anthropology carried an aura of objectivity and precision, of proper science. It helped, too, that rules were drawn up for the standardizing of anthropological measurements, at the International Congress of Anthropology in Monaco in 1906 and at Geneva in 1912. Often devoid of theory and focusing on preconceived human types, passionate skeletal studies and catalogs continued in some contexts throughout the twentieth century. Sometimes the goal was partly to understand particular medical conditions evident in the skeletal structures of particular populations. From the beginning of national projects, with the emergence of nation states and empires, phenotypic markers played important roles in identifying and demarcating social groups, and in establishing criteria of citizenship (Lindee and Ventura Santos 2012). While national projects were focused on the broad and apparently neutral and culture-free documentation of "difference" and "variation," they usually separated "us" from "them," expressing ethnocentric concerns with race and purity.

In her early historical account of the instruments of physical anthropology, Hoyme (1953) had no doubts about the bright future ahead. Speculating about the next fifty years, at the time of the discovery of the double helix, she suggested:

> (it is) fairly certain . . . that anthropometry, the hallmark of physical anthropology, will be retained among the techniques of the anthropologist, although techniques from other biological sciences will undoubtedly be adopted where needed . . . There are many problems involving size and proportion not yet solved, and perhaps not yet suggested by genetics, physiology, ecology, and comparative anatomy, for which anthropometry can provide basic data. (Hoyme 1953: 425)

Yet, modern readers cannot help but wonder why people bothered with all the measurements and indices produced. Hoyme (1953: 423) admitted that it "is somewhat . . . difficult to draw conclusions as to why these measurements and indices were considered worthy of investigation," adding that "it is necessary to examine the studies against the climate of opinion of the period in which they were carried out." One of their main justifications was their role in sustaining national projects, establishing variation, maintaining identities and boundaries. Bodies, like biology, remained static while history was dynamic.

There was an important counter voice, however. Franz Boas, who studied the cranial form (the cephalic index or the ratio of the length to the width of the head) of immigrants and their children living in New York, argued that adult phenotypes were dependent upon environmental exposures during human development; head shapes converged to a common type due to similar environmental pressures, and, as a result, he concluded, "we are necessarily led to grant . . . the great plasticity of the mental make-up of human types" (Boas 1911: 64–65). Presented as a critique of biological determinism and the thought style of racial classification, Boas's work was controversial and largely ignored. As Stephen Jay Gould (1996: 58) observed in *The Mismeasure of Man*, "[s]cientists are used to analyzing the data of their peers, but few are sufficiently interested in history to apply the method to their predecessors." Recent reanalysis of Boas's results, applying analytical techniques not available to him, provides strong support for his claim about the plasticity of human biology (Gravlee, Bernard, and Leonard 2003). Perhaps this signaled an early, and somewhat premature, social turn in physical anthropology.

Modern genetics was born at the beginning of the twentieth century with the rediscovery of Mendel's laws of inheritance. By the middle of the century, a series of critical developments revealed the molecular mechanisms involved. Twentieth-century advances in cell research also made it possible to extract, store, and manipulate cellular material outside the host organism. With the "modern evolutionary synthesis" of the 1940s and the "new genetics" of the 1960s onwards, biological anthropology took off. Theoretical and technical innovations allowed geneticists and scholars in related fields to explore the contours and structures of genetic material, to "voyage" into the "universe within," in the language of the Human Genome Project in 2000.

It was not until the 1990s, however, that DNA began to replace skeletal material and other sources (texts, languages, and archaeological remains) as *the* avenue into human history and variation. The study of phenotypic variation (skin color, skull shapes, hair texture, fingerprints, and so on) that for decades was seen as the hallmark of physical anthropology no longer had center stage. A few geneticists and biological anthropologists began to see bones as almost trivial antiques in the history of studies of human variation (see Sommer 2008), somewhat like ancient manuscripts, secondary to DNA sequences and gene frequencies. Keeping in mind, however, that formerly discarded sources of data sometimes acquire new meaning as a result of conceptual and methodological advances (ancient DNA being one pertinent example), skeletal morphology should not be discounted.

The new genetics necessarily redefined the conceptual landscape and biometrics of human difference. Genotypic differences seem not to correspond or submit to earlier framings of phenotypic differences and classifications. Many biological anthropologists, in fact, have sought to change popular understandings of race, emphasizing that race has no biological basis. This does not, however, rule out the possibility of somehow reintroducing earlier categorizations of races under the modern labels of genotypes, populations, and ethnic groups (Haraway 1996, Bliss 2012).

As genomics acquired a central place in national projects, as a powerful avenue into variation and history, the issue of the genetics of race and health became an important concern. National collections and genetic databases drew upon a series of developments in the history of states and bureaucracies, including practices of writing, accounting, and monitoring. Developed in one nation state after another, statistics represented biopolitical concerns with monitoring and governing the health of the national body. Health statistics, in a sense, establish the birthmarks of the national body. One of the recent territories of biomedicine is the move beyond "national" genomes to comparative datasets such as that of the International Haplotype Map Project and the Human Genome Diversity Project, building on the power of genome-wide scans and related technological and methodological advances. Drawing upon its comparative tool kits, anthropology has often been involved in such projects, sometimes as a key player (M'charek 2005, Reardon 2005).

Just as the anthropometry of past centuries was moved from the center of national projects and studies of human variation, the focus on "human" genomics of recent decades is bound to change. While genomic avenues into history, difference, and relatedness have proved quite powerful, it seems likely that they, too, will be redirected. Already, a series of developments can be observed on the horizon, among them the escalation of computing power with advances in bioinformatics and computer technology facilitating complete genome sequencing at a low price, the growth of personal genomics and digital networking, and new forms of biopolitics and governance (Prainsack, Schicktanz, and Werner-Felmayer 2014).

One advancing empirical issue for students of human genome variability is that of the stability and temporality of the genome. While there is a tendency to assume that the genetic signatures of a person are present in every cell in the body and, moreover, the same throughout the life course, this does not seem to be the case (Barnes and Dupré 2008: 76). The same body may exhibit different signatures and, moreover, the signatures may change during the lifetime. One reason is that of gestational chimerism, the intermingling during pregnancy of cells from two or more genetically distinct bodies. In the orthodox view of human reproduction, the human foetus and the mother are two distinct individuals with distinct evolutionary objectives, maintaining the status of "all self" during the process of gestation. It seems that a woman who has been pregnant is likely to be a chimera, acquiring "foreign" cells and genomes from every pregnancy. Her body, in other words, is potentially inscribed with each pregnancy, whether or not it resulted in birth, and hence the presence of some genetic material from the father in her body. This highlights troubled body boundaries or interfaces between sexual partners, siblings, and across generations (Dupré 2010, Martin 2010, Kelly 2012). Human variation, then, in genomic terms, is not as fixed and stable an epistemic field as often assumed. The "book of life," given the rhetoric developed in the course of the mapping of the human genome, seems unstable, inconsistent, and poorly edited. Clearly, the genome has a "social" aspect, embodying context and historical practices, much like phenotypic markers such as cephalic indexes and skin color.

6.2 The relational superhuman

Boas's conclusions about environmental influences on cephalic proportions were debated and eventually more or less forgotten. Recently, however, in the wake of developments in epigenetics, they have been brought back to the agenda and reconsidered (Lock 2015). Environmental exposures, it is argued, become *literally* embodied, with consequences for health and illness, possibly inherited over generations (Landecker and Panofsky 2013). The lives of our parents and ancestors, in other words, and the traditions and conditions of their communities in all their complexity, from dietary factors and exposure to toxic substances to behavioral habits, are embodied and memorized in our genomes, turning on some genes and silencing others, leaving a lasting "hereditary" impact in a somewhat neo-Lamarckian fashion. Given the growing evidence of epigenetics, the body is inevitably entangled, firmly connected to the environment.

Another form of entanglement is that of microbiomes. The human genome, it turns out, is fundamentally conflated with the genomes of other organisms, mainly in the guts. It seems likely that studies of the human microbiome will profoundly affect anthropology (Benezra, DeStephano, and Gordon 2012). Microbes are partly vertically transmitted, between people more or less belonging to the same community, and, as a result, knowledge of their variation and history enhances understanding of population ancestry in the conventional sense, tracking the migrations of individuals and populations. The entangled view, on the other hand, seems to necessitate reconsideration of current thinking with populations. The evidence already accumulated necessarily destabilizes the issue of human variation and national genomes.

The fact that what is normally considered the "human" genome is fundamentally mixed with the genomes of other organisms complicates the picture of human variation. Anthropologists, it seems, need to qualify their language about the microbiome as "more-than-human," as an addition to a host, independent of human variation. Not only is the microbiome mostly inside our bodies, in a fundamental sense it is "us." Keeping in mind that the microbiome plays an important role for our immune system and that it is often used to separate self and non-self, it is "difficult to characterize our symbiotic partners as entirely distinct from ourselves" (Dupré 2010: 28). The geopolitical metaphors of "invasion," "foreign" entities, and bodies "at war," one may add, common in immunology, seem to be on the defense (Martin 2010).

If the microbiome is "us," the human genome that has been heralded as the key to practically everything human is only a fraction of the whole picture. It seems likely that casting the genomic net more widely to incorporate the vast terrain of microbiomes, vastly expanding the range to explore, is bound to radically change current notions of both *Homo sapiens* and human variation. The first results of this trend, based on a very small sample, indicated that while the composition of the microbiota within the guts of different individuals is distinct they cluster as three "enterotypes." These enterotypes, it seemed, did not correlate with classical categorizations based on "race," color, morphology, and continental background (Arumugam *et al.* 2011). Some, however, warn that because of the relentless change of the microbiota due to modification of lifestyle and diet categorization may not be that useful. Also, it would make more sense to speak of a "continuum or gradient of species functionality rather than discontinuous variation with segregated types" (Jeffrey, Claesson, and O'Toole 2012: 592). This is an argument that echoes debates on the clustering and geography of "human" genomes in the narrower sense (Serre and Pääbo 2004). Microbiomes, no doubt, will be an important issue in the coming years. Anthropologists should attend to the unfolding discussion and contribute, with their perspectives on human variation and becoming.

Drawing upon research on the microbiome, some scholars suggest that the common notions of "gut feelings" and "gut reactions" should not be simply seen as metaphors but rather as indications of a post-humanist "gut sociality":

> While I may pay little attention to the intestinal flora that colonize my gut, I am co-implicated with these more-than-human embodied others in various intra- and extracorporeal circuits of sociality ... Gut sociality thus problematizes the idea that sociality would be solely an *interhuman* mode of relation, and instead suggests a notion of sociality that is posthumanist in orientation; this understanding opens up the term "sociality" beyond its common privileging of the human subject. (Neimanis 2013: 218; emphasis in the original)

To some extent, the idea of the relational supraorganism envisioned by studies of chimera, microbiomes, and epigenetic processes is echoed by Amazonian and Melanesian ideas about chronically unstable bodies. Vilaça (2005: 459) suggests that many Amazonian ethnographies "challenge the understanding of the body founded on a biological substrate," since "substances transmitted through conception are less important than or at most equally important to those acquired and exchanged through social practice." Similarly, Strathern (2004: 11) argues, in the context of Melanesian ethnography, that the whole "dividual" person, "as part of a nexus of relationships," is only "completed by relationships with others." This also resonates with pre-Enlightenment European views of porous bodies, open to influence through social practice.

What would Charles Darwin, one may wonder, have made of recent studies of human microbiomes? Would they testify to porous dividual kinds of beings with continually becoming varieties? The questions may be more relevant than one might think. During his *Beagle* voyage in 1835, Darwin was often preoccupied with human variation in the tropics: "Whence have these people come from?," he wondered. While he was puzzled by the bodily mixtures he saw, his language was remarkably devoid of derogatory racial judgment. His encounter with the inhabitants of the Chiloè Islands off the coast of Chile prompted interesting questions about race and species, humans and parasites. The large, black lice that he collected were different from the English lice he knew and, he was told, they died promptly when crawling on British sailors. Different species of lice seemed to adapt to different races of humans. "Man springing from one stock according his *varieties* having different parasites," he observed, probably reasoning that lice, just as humans, diverged from a common ancestor (see Desmond and Moore 2009: 98). Then he bluntly stated: "It leads one into many reflections." Later he speculated if this could mean that humans were different species. Unlike many of his British contemporaries, however, he concluded this was not the case, producing evidence that different species of parasite can infect the same race and that minor changes in the host body can repel its parasites (Desmond and Moore 2009: 389, n. 57).

6.3 Human variation after the biosocial turn

Inevitably, the discipline of anthropology is part of the ongoing negotiation of the social and the biological, the biosocial turn, given its conventional fragmenting of *Anthropos* under the labels of bodies and societies. This is shown, as I have argued, by the central notion of human variation and differences. Indeed, it may represent one of the cutting edges of the current biosocial turn. Students of human variation – of skeletal material, genomes, as well as microbiomes – need to attend to biosocial relations and body worlds.

While the measurements of the bone people during the last century were exceedingly detailed, the classificatory scheme within which they were placed was usually highly simplistic and archaic, with an implicit racial tone. Inevitably, the arrival of human genetics and biological anthropology after World War II provided a new avenue into the understanding of human variability, moving the kinds of measurements and classifications of skeletal material typically practiced by physical anthropologists during the first half of the twentieth century to the side-lines. Now, as the evidence on the epigenome, chimera, and polygenomic humans unfolds, the issue needs to be problematized and reframed once again. While studies of both skeletal material and genomics will continue to revise the story of human differences and history, national projects and the sciences of human variation – in particular, genomics and anthropology – are bound to undergo conceptual change.

Observations on human variation, of course, are based on measurements at a particular time and place, capturing snapshots of human life at high speed. Whatever the definition of human "biology" adhered to, the *assembly* of such snapshots should be informed by a meaningful theory of variation. Although observers of variation sometimes have an intuition about what matters and some of their snapshots later generate unexpected insights, the old positivist "problem orientation" of many physical anthropologists – namely, "amass enough data and problems will sort themselves out" (Scott *et al.* 2000: 347) – will not do. The anthropology of variation would benefit from attending to the nature of body worlds and the biosocial realities of variation, moving beyond material fetishes and geneticized bodies.

Also, when *reading* available snapshots it is important to keep in mind the relational and the temporal. As Gare (1995: 107) argues in the context of nature/culture debates, it is important "to develop a science which takes becoming as basic . . . and conceives 'beings' as islands of stability within the flux of becoming." Social and cultural anthropologists have often referred to their observations from the field as the "ethnographic present," acknowledging the limits of their field of vision and emphasizing the need to accommodate the flux and connections of becoming. Similarly, it makes sense, for the purpose of capturing the moment of becoming in the flux of human life, to speak of human variation as a moving "biosocial present," as shifting "local biologies" in Lock and Kaufert's sense (2001), continually situated in a dynamic context. This is not just an academic, anthropological exercise; it has profound implications for the understanding of health and well-being, justice and inequality – for meaningful future biopolitics.

7 Nim Chimpsky *et al.*: human–animal relations

Terrific achievement. To cast off apehood in five years and gallop through the whole evolution of mankind!

Franz Kafka, "A Report to the Academy," 1917[1]

Chimpanzees seem to occupy a special position in recent writings on the nature/culture divide, as a liminal species at the main border of modernist discourse. Partly drawing upon the life and work of Nim Chimpsky (1973–2000), a chimpanzee raised in experimental and familial settings in the United States in order to test hypotheses about innate and acquired mental capacities, especially language, this chapter discusses the history of comparisons of chimpanzees and people and, more broadly, the relations of humans and other species. If one takes Chimpsky's near-namesake Noam Chomsky seriously, assuming that language as we know it rests on an innate language "device," one is inclined to ask what such a device consists of, how it developed, and what might be learned through comparisons of humans, other primates, and other "lower" species, an issue only recently addressed by Chomsky himself (see Hauser, Chomsky, and Fitch 2002).

I shall argue that academic debates about language and mind generated by Chimpsky, other chimpanzees, and their human and non-human collaborators reflect different understandings of the nature/culture divide and what used to be called the animal kingdom. While experiments with the language and sociality of chimpanzees and other species are often non-conclusive and sometimes misguided, they usually bring home important points about ourselves and our relations to other species. The outline of the chapter is as follows: I first briefly discuss the history of human experimenting with chimpanzees, partly with reference to an illuminating and perceptive short story by Franz Kafka that seems to have anticipated what was to come. I then discuss the amazing case of Nim Chimpsky and some of the lessons we may and may

[1] Excerpt from *The Complete Stories* by Franz Kafka, copyright © 1946, 1947, 1948, 1954, 1958, 1971 by Schocken Books, an imprint of the Knopf Doubleday Group, a division of Random House LLC. Used by permission of Schocken Books, an imprint of the Knopf Doubleday Publishing Group, a division of Penguin Random House LLC. All rights reserved.

not learn from it. This is followed by a brief tour into evolutionary and comparative discourse on primate becomings and capacities and linguistic doubts about language universals. Finally, I discuss the growing role of nonhuman animals in language experiments, summing up the arguments in the chapter and their implications for discussions of the nature/society divide and anthropological engagement with it.

Chimpanzees continue to attract our attention, illuminating the human–animal condition one way or another. Not only is Nim the subject of both a recent biography, *Nim Chimpsky: The Chimp Who Would Be Human* (Hess 2008), and a documentary, *Project Nim* (2011), by James Marsh, also now there is an "autobiography" of a chimpanzee, *Me Cheeta: The Autobiography* (Lever 2008), nominated for the Booker Man Prize in 2009. Moreover, one of the most popular films of 2011 was the science-fiction film *Rise of the Planet of the Apes*, a restart of an earlier popular series, featuring a chimpanzee named Caesar who is a test subject for a new Alzheimer's cure and, as a result, becomes highly intelligent and eventually leads a revolt of the lab primates. Time and again, animal experiments and their accounts in all kinds of media reveal our own preoccupations and the ways in which we relate to other animals.

Relations between humans and other animals, almost by definition, especially perhaps relations between humans and other great apes, represent a central theme in both anthropology and philosophy, inviting fundamental questions about us and them, nature and culture. Both fields, however, have tended to take a highly anthropocentric position. Ingold (1988: 1) has indicated that the comparative anthropological project itself might be the problem: "Does not the anthropological project of cross-cultural comparison rest upon an implicit assumption of human uniqueness *vis-à-vis* other animals that is fundamentally anthropocentric?" Derrida (2008: 32) launched a similar critique of philosophy, emphasizing the tendency to discuss "the animal" in the singular, almost invariably in contradiction to "the human": "All the philosophers we will investigate (from Aristotle to Lacan . . .), all of them say the same thing: the animal is deprived of language. Or, more precisely, of response, of a response that could be precisely and rigorously distinguished from a reaction." Within this general singularity, Derrida continues, "are *all the living things* that man does not recognize as his fellows, his neighbors, or his brothers. And that is so in spite of all the infinite space that separates the lizard from the dog, . . . the parrot from the chimpanzee" (2008: 34; emphasis in the original). Interestingly, however, while Derrida (2008: 99) emphasizes "structural difference between nonhuman types of animal" – presumably he is referring to biological differences between species – he seems to see no point in exploring the ethnography of human takes on relations with other animals, restricting his own account to biblical narrative, ancient Greece, and a handful of twentieth-century continental European philosophers.

While Derrida's dismissal of ethnography may resonate with his troubled relationship with anthropology (Morris 2007), it seems to contradict his emphasis on pluralizing voice. In fact, he might have found much inspiring thought in recent ethnographies, in particular the broad kind of "multispecies ethnography" (Kirksey and Helmreich 2010) that seeks to go beyond the traditional scale of primates with which many biologically minded anthropologists operate. A rapprochement of anthropology and philosophy on this score is timely and urgent. Recently, the field of animal studies has been vastly expanded and revitalized through new avenues, including those of cognitive studies, primatology, and the animal rights movement (see, for example, Haraway 2008, Kirksey and Helmreich 2010). The revival of totemism, the reappearance and escalation of animal sensibilities in spite of Lévi-Strauss's dismissal of the "totemic illusion," seems to continue on full course. More generally, there is a growing body of work on both the labor of animals and the traffic of animal substances. Harré (2009: xii) has called for "a comprehensive view of how organic beings and their parts and remains have actually been used, together with some idea of the kind of people who have used them and for what scientific purposes." This means paying attention to and granting agency to all those participating in the "living laboratory," including insects, organic clocks, model organisms, and language-learning chimpanzees. "What if a chimpanzee," Harré (2009: 10) asks, "has its own agenda in interacting with those who are studying it?"

7.1 Reports to the academy: almost human, almost chimpanzee

In popular as well as academic discourse, chimpanzees are often seen as a liminal species, endowed with cognitive capacities and socialities that almost elevate them above nature, nearly making them human, almost possessing culture. Humans, likewise, are frequently presented as apelike beings, quasi-chimpanzees, firmly rooted in their primate past despite their cultural heritage. This is exemplified by popular book titles such as *Almost Chimpanzee* (Cohen 2010). Recent developments in genomics have highlighted the same themes, in new terms – thus, *What It Means to Be 98% Chimpanzee: Apes, People, and Their Genes* (Marks 2002). While the popular qualifying reference to "almost" highlights a categorical difference between nature and culture, establishing a narrow but fertile comparative zone of meaningful difference for both biological and social anthropology, it also suggests fascination with experimenting and transcendence – in particular, as we will see, the possibility of communication across the species divide – expanding at the same time the community of "culture" to other primates and possibly beyond. The etymology of the word "chimpanzee," introduced into the vocabulary of the Western world in 1738, is significant. Derived from *kivili-chimpenze*, a term in the Tshiluba language spoken in what is now the Democratic Republic of Congo, it translates as "mock human," suggesting perhaps that Tshiluba speakers may have seen chimpanzees as the product of couplings between humans and other species. The first ape that became widely known in Europe arrived on an English merchant ship, the *Speaker*, in London in 1738. Coming from Angola, she was reported as "an animal of remarkably and terribly hideous countenance ... called by the name Chimpanzee" (see Marks 2002: 19).

Despite their depressing early impression, chimpanzees eventually became the subject of systematic experimenting and observation, sometimes in close company to humans. A widely discussed case was that of Köhler's "Anthropoid Station" on the island of Tenerife between 1912 and 1920. In his account on the experiments to the Prussian Academy, Köhler (1956) demonstrated that chimpanzees were capable of solving simple practical problems, typically by rearranging boxes as tools for getting access to food, as if they were making a ladder or, as Köhler reasoned, climbing a tree. While anthropoid apes, contrary to common assumptions, turned out to be capable of creating and using tools, the absence of the sign from their world of communication seemed to reaffirm a boundary between us and them – exemplifying what Haraway (1989: 10) calls "simian Orientalism." Köhler's historic experimenting at the Anthropoid Station and its reporting to the Prussian Academy may have inspired Kafka to write his much-cited 1917 short story "A Report to the Academy" about Red Peter, a chimpanzee performing on the variety stage who turned human and told his story. I return to Kafka's motives later on. In any case, the saga of Red Peter took on a life of its own. Not only did it foreshadow recent discussions of the use and treatment of laboratory animals, it also anticipated language experiments with chimpanzees, including Nim Chimpsky, and the idea of chimp-human transcendence, beyond the almost-culture rhetoric of beyond-nature.

Kafka's story is centered on a chimpanzee who is captured on the Gold Coast of Africa and five years later, following a traumatic experience, is able to deliver an account of his life: "Honored members of the Academy! You have done me the honor of inviting me to give your Academy an account of the life I formerly lead as an ape" (Kafka 1983: 250). After capture, when he came to his senses, Red Peter found himself inside a cage: "The whole construction was too low for me to stand up in and too narrow to sit down in ... Hopelessly sobbing, painfully hunting for fleas, apathetically licking a cocoanut, beating my skull against the locker, sticking out my tongue at anyone who came near me ... But over and above it all only one feeling: no way out" (Kafka 1983: 252–253).

Red Peter offers an account of how he acquired language. One evening he took hold of a schnapps bottle and "like a professional drinker, with rolling eyes and full throat, actually and truly drank it empty," and in the heat of the moment "called a brief and unmistakable 'Hallo!' breaking into human speech, and with this outburst broke into the human community, and felt its echo: 'Listen, he's talking!' like a caress over the whole of my sweat-drenched body" (Kafka 1983: 258). At the same time, he offers some witty observations regarding his spectacular transition to humanity. Breaking into the human community by means of speech meant bypassing the evolutionary past, transcending the nature of the ape:

> The strong wind that blew after me out of my past began to slacken; today it is only a gentle puff of air that plays around my heels ... To put it plainly, much as I like expressing myself in images, to put it plainly: your life as apes, gentlemen, insofar as something of that kind lies behind you, cannot be farther removed from you than mine is from me. Yet everyone on earth feels a tickling at the heels; the small chimpanzee and the great Achilles alike. (Kafka 1983: 251)

"And so I learned things, gentlemen," Red Peter concludes, "My ape nature fled out of me ... With an effort which up till now has never been repeated I managed to reach the cultural level of an average European" (Kafka 1983: 258). Those listening to Red Peter are impressed by his account and his accomplishments. One of them observes (Kafka 1983: 261): "Terrific achievement. To cast off apehood in five years and gallop through the whole evolution of mankind!"

Why would Kafka write such a narrative? As already mentioned, he may have been inspired by Köhler's account of the chimpanzees in Tenerife. There must be more to the story, however. Coetzee (1999: 15) speculates, partly through his fictive character Elizabeth Costello, a moral philosopher and animal rights activist, that the saga of Red Peter on the variety stage may have been "an allegory of Kafka the Jew performing for the Gentiles." Coetzee (1999: 32) has Costello reason: "If Red Peter took it upon himself to make an arduous descent from the silence of the beasts to the gabble of reason in the spirit of the scapegoat, the chosen one, then his amanuensis was a scapegoat from birth, with a presentiment, a *Vorgefühl*, for the massacre of the chosen people that was to take place soon after his death." Kafka, Costello (1999: 40) suggests, saw both himself and Red Peter "as hybrids, as monstrous thinking devices mounted inexplicably on suffering animal bodies." Levinas similarly juxtaposes chimpanzees and Jewish prisoners of war under the Nazi regime: "We were subhuman, a gang of apes" (quoted in Derrida 2008: 117). Derrida comments that the animal "remains for Levinas what it will have been for the whole Cartesian-type tradition: a machine that doesn't speak, that doesn't have access to sense, that can at best imitate 'signifiers without a signified' ..., *a sort of monkey with 'monkey talk'*" (2008: 117; emphasis added).

Whatever the concerns or motives that drove Kafka to write the story, the plot seems to anticipate several highly modern issues – the case of Nim Chimpsky, the biography of Cheeta, the expanding animal rights movement, the unprecedented subjugation of the animal, the biopolitics of life itself, and increased attention to interspecies relations in social theory – all of which figure prominently in the unfolding and collapse of the nature/society divide.

7.2 The life and work of Nim Chimpsky

Serious language experimenting with chimpanzees began in the 1950s. The Hayes tried to teach "Vicki" to speak, raising her as if she were a human child. Apparently she only learned four words in six years ("mama," "papa," "cup," and "up"). Later, the Gardners taught Washoe, caught in the wild in 1966, gestural language, realizing the inadequacies of chimpanzees' vocal chords for forming words. Washoe accumulated 150 signs in five years, appearing to make sentences and invent new combinations like "candy fruit" (for "watermelon"). The study of Nim Chimpsky signaled a brave new era of ape research and language experimenting, with intense socializing, this time in the wilderness of hippie Manhattan. Sometimes, however, the goal of rigorous testing and observation was overshadowed by a chaotic atmosphere and frequent changes in personnel.

Nim was born on November 19, 1973 in Norman, Oklahoma. His mother was Carolyn, a wild-born eighteen-year-old, imported from Africa as an infant and sold to Chicago Zoological Society. The same year Nim was born, Geertz (1973: 66) summed up attempts to distinguish between species in terms of difference in kind rather than in degree: "Man can talk, can symbolize, can acquire culture, this argument goes, but the chimpanzee (and, by extension, all less-endowed animals) cannot. Therefore, man is unique in this regard." Geertz's qualification "this argument goes" seems to suggest that statements about human uniqueness had been elevated to clichés; such attempts, he argued metaphorically, "see adulthood as a sudden transformation of childhood and miss adolescence altogether" (1973: 66)." Devoid of sign making, non-human animals – our presumed closest relatives chimpanzees – were mere signifiers. Nim would become a test case, potentially resolving the issue on the language device highlighted by a famous debate between Skinner and Chomsky (Chomsky 1959) – whence the name "Nim Chimpsky."

The experimental subject began his stormy life as "number 37" in primate records. About two weeks old Nim was sent to New York, to join a human family and to participate in a major ape language study at Columbia University, Project Nim, led by psychologist Herbert S. Terrace. Terrace's aim (1987) was to raise the chimpanzee as if he were a human child and to explore whether he could learn to use American Sign Language (ASL). This would be a challenge to the thesis often attributed to Chomsky that language is inherent and unique for humans. As Nim's biographer remarks, "Like race-horses breathlessly running around the track, the chimps and the psychologists who were training them were headed toward an elusive finishing line, where Skinner and Chomsky – each with an entourage of followers – stood, waiting to judge the race" (Hess 2008: 132). While thousands of primates (monkeys and apes) lived in laboratories during the 1980s, none of them received the same attention that was shown to Nim. He became a public sensation, frequently discussed in the media. At one point anthropologist-novelist Kurt Vonnegut visited him.

At first Nim was adopted and raised by psychology student Stephanie LaFarge. She would nurse him in her family apartment along with her own children; for a while Nim requested breast-feeding from his "mother," and LaFarge's daughter Jenny once asked if Nim was a new baby or a new pet. At Columbia, Nim's linguistic performance and abilities were subjected to numerous inscription devices, translated into tables and graphs. While few of the instructors were fluent in sign language, Nim turned out to be a real signer. One of the dialogues he had was as follows: "What do you want to do?" / "Rick tickle Nim" / "Tickle where?" / "Tickle here" (see Anderson 2004: 278). In the course of his life, Nim encountered a series of nurses, teachers, labs, and experiments – engaging several families, research teams, and universities in a somewhat unique living laboratory. Most of the people who came anywhere near him developed a personal relationship with him; Nim would play a joke on them or teach them a sign. No one, however, wanted to be stuck with a chimpanzee for good. When the project ran out of funds and Nim's family, friends, and bosses had turned their attention to other things, he was sold to a medical lab to be the subject of experiments. Eventually, thanks to the intervention of some of his human friends, he was placed in a sanctuary in Texas. There he showed signs of depression, but later he settled in with the inmates, gradually assuming a central place: "He watched the other animals – elephants, giraffes, monkeys, and gibbons – from a captain's walk in his cage as they arrived at the ranch" (Hess 2008: 320). On March 10, 2000, he had a massive heart attack in the middle of playing. A few days later, a small group gathered for a memorial. Only one representative from Project Nim showed up, Stephanie LaFarge.

Many of those who followed accounts of the language experiments at Columbia in the media and the literature expected that Terrace and his colleagues would proudly announce that Nim had become the first ape to demonstrate a human kind of language structure. However, Terrace asserted to the contrary in a *Science* article (Terrace *et al.* 1979) that there was no evidence that Nim had any ability to string signs together in the manner of human language. Marks (2002: 182) sums up the results of sign-language experiments with apes: "First, they do have the capacity to manipulate a symbolic system given to them by humans, and to communicate with it. Second, unfortunately, they have nothing to say. And third, they do not use any such system in the wild." Language, Marks (2002: 184) goes on, "is just not a chimpanzee thing." Terrace's conclusions, however, may have been biased against Nim, complicated by other concerns: "Was it possible that Terrace had taken such a negative view of Nim's abilities because dumping a language-using, humanized ape back in a cage with non-language-using chimpanzees would be worse than doing that to an animal without the ability to use language?" (Singer 2011; see also Terrace 2011). Significantly, some of Nim's caregivers were surprised by Terrace's apparent surrender – his "change of mind," to use his own words. After all, Nim would make sentences much like a human child, initiating conversations that could hardly be pure imitation. Moreover, to some he showed signs of self-awareness and strong emotional bonding. Other studies, notably Savage-Rumbaugh's keyboard signing with the bonobo (pygmy chimpanzee) "Kanzi" (Savage-Rumbaugh and Lewin 1994), born in 1980, seem to have reopened the issue of the uniqueness of human language.

Despite all the experimenting with apes and language, it has been difficult to settle what was taken as a Chomsky–Skinner issue. But perhaps both sides of the debate missed the point. Recently, the intellectual front-line has moved from the innate language device to other issues, to experimenting among chimpanzees with problem solving and prosocial behavior, studies of the evolution of social interaction, genetic analyses, phylogenies of language, and doubts about language universals – to mention a few related themes.

7.3 Becoming primates, becoming human

Early reflections on chimpanzee communication tended to emphasize the role of imitation in vocal signaling, its limits, and its developmental and evolutionary implications. If human language was not simply the gift of God, why and how did it emerge and what, if anything, might be learned from chimpanzees and other close relatives? These were central and heated issues in nineteenth-century discussions, involving, among others, C. Darwin, M.M. Müller, E.B. Tylor, and R.L. Garner. In *The Descent of Man* Darwin suggested that imitation might have been the forerunner to the language of "barbarous races" and, eventually, civilization:

> The strong tendency in our nearest allies, the monkeys, in microcephalous idiots, and in the barbarous races of mankind, to imitate whatever they hear deserves notice. As monkeys certainly understand much that is said to them by man, and as in a state of nature they utter signal-cries of danger to their fellows, it does not appear altogether incredible, that some unusually wise apelike animal should have thought of imitating

the growl of a beast of prey, so as to indicate to his fellow monkeys the nature of the expected danger. And this would have been a first step in the formation of language. (Darwin 2008: 239)

Such ideas seemed to draw upon fairly common perceptions of higher apes as masters of mimicry. Indeed, many languages have words related to "aping" for the act of copying.

At the beginning of the twentieth century, evolutionary speculations about the mentality of primates gave rise to behavioral, psychological experimenting. In their book *Ape, Primitive Man and Child: Essays in the History of Behavior* Luria and Vygotsky extended evolutionary reasoning to ontogenetic theory and observation, maintaining the tripartite distinctions made earlier by Darwin, Köhler, and others about the mentality of chimpanzees, primitives, and humans: "The absence of even the rudiments of speech, in the broadest sense of the term – the ability to make a sign, to introduce auxiliary psychological resources – that everywhere distinguishes human behavior and culture, is what draws the boundary between the apes and the most primitive man" (Luria and Vygotsky 1992: 31). This picture was radically changed through numerous studies of behavior and tool use among a variety of non-human animals. New insights were generated, for instance, by detailed field studies among chimpanzees in the wild pioneered by Goodall's book *In the Shadow of Man* (1971), based on participant observations in Tanzania in the 1960s. Another important chimpanzee study was that of Kortlandt (1986), who explored tool use in Guinea and Liberia. His evidence suggested that chimpanzee communities had copied habits of cracking oil-palm nuts by means of stone tools from local farmers after careful observation. These communities, Kortlandt suggested (1986: 77), "may represent the first identifiable cases of direct cultural transmission of technology from man to animal in the wild."

Over recent years, research in several fields, including anthropology, genomics, archaeology, primatology, cognitive psychology, and linguistics, has redefined the historiography of humans, their ancestors, closest relatives, and means of communication (see, for instance, Gibson and Ingold 1993, Marks 2002). At the same time, claims about what counts as grand narrative have changed. In particular, with the development of the "modern evolutionary synthesis" of the 1940s and the "new genetics" of the 1960s onwards (Palsson 2007), skeletal material was moved from the center towards the margin. Some biological anthropologists began to see bones as secondary to DNA sequences and gene frequencies. There is broad agreement on some things – including rough timelines, evolutionary trajectories, and biological and behavioral differences – although many important theoretical issues remain unsettled and debated.

About six million years ago, a population of African apes separated into two distinct species, eventually leading to humans and chimpanzees (see, for instance, Cavalli-Sforza 2000: 57). *Homo sapiens* arrived on the scene about 500,000 years ago and modern human language between 50,000 and 150,000 years ago. One of the evolving narratives on the evolution of language in the wake of the new genetics is that of the "Forkhead box P2" or *FOXP2* genes. Recently it was reported that the language problem associated with the so-called KE family in Britain, a partially compromised ability to speak and process words, was linked to a mutation in one of their two *FOXP2* genes (Lai *et al.* 2001). Almost immediately, this discovery sparked some evolutionary speculations. Human *FOXP2*, at most 200,000 years old, it was argued, signaled a selective sweep paving the way for modern language (Krause *et al.* 2007). The two amino changes involved in humans seem to have occurred after they separated from chimpanzees.

Neandertals, who split off from modern humans more than half a million years ago, turn out to have the same two changes in their *FOXP2* as modern humans. Their case may be particularly interesting, due to their close phylogenetic affinity to modern humans, but here the only evidence available is that of skeletal material. It is only very recently that a Neandertal genome has been sequenced, by Svante Pääbo and colleagues. It now seems, given Pääbo's evidence, that humans and Neandertals were not only contemporaries for a long time, they also interbred (see Kolbert 2011). These developments signaled a kind of truce in debates between biological and physical anthropologists about what kinds of material provide the best data about the hominid past; skeletal material was *still* useful – albeit on limited terms – as the source of DNA. Had Neandertals or other relatives closer to us than chimpanzees survived to the present day and we were in a position to interact with them next door or "in the wild," our fascination with human exceptionalism and the capacity for language might have been subdued – and chimpanzees would probably be far less appealing as experimental objects and liminal icons illuminating the nature/society divide. Would we simply attribute liminality to another species, drawing the lines with higher resolution, focusing on Neandertals, applying the Tshiluba notion of "mock human" (*kivili-chimpenze*) – or some equivalent – to Neandertals rather than chimpanzees? Or would we perhaps take the existence of Neandertals as evidence of continuity, representing "adolescence," to use Geertz's analogy mentioned above, the smooth transition from childhood to adulthood in hominid evolution? Would we, in other words, have relaxed our assumptions about human uniqueness? But, then, evolution and history would have unfolded differently and we have no way of knowing who would be "us" in "our" accounts, or how.

Some geneticists and cognitivists imagine the *FOXP2* saga has established order in the house, finally testifying to an innate language faculty, attributing it to a couple of genes. The jury, however, is still out. Other developments – in particular, the emerging evidence of epigenetics – suggest the roles of DNA and the connections between genotype and phenotype are not only highly complex but also non-deterministic (see, for example, Moss 2003). Perhaps when reflecting on the issue of the genetic roots of language we might keep in mind an earlier moment in the history of experimenting with chimpanzees: Just as the Gardners realized that despite her inadequate vocal chords Washoe might not only learn to "speak" by means of sign language but also to teach her adopted son some signs, should we not consider the possibility that humans are capable of language whatever the mutations on the *FOXP2* genes? Genetic avenues into history and relatedness have undoubtedly proved quite powerful, but they are likely to be redirected in the future with the growing destabilization of gene talk – much like the single-minded bone talk of the past with its cephalic index and measurements.

We might also keep in mind that the broad quest for language has not only been extended to other primates but also to many other far more distant relatives, including whales and dolphins (Rendell and Whitehead 2001) and even "lower forms of life" such as honeybees (Crist 2004). The dance of the honeybee, Crist (2004: 35) observes, disturbs "'the great chain of being' still at large . . .; the picture of man (and other 'higher mammals') at the apex and invertebrates in the basement of a hierarchy of ability and value." While many people probably find it mind-blowing, to say the least, to attribute mind and language to insects, creatures way below primates and mammals in the cladograms of evolutionary discourse, it has been notoriously difficult to exclude honeybees from the republic on fair and objective grounds (for another view, see Anderson 2004).

7.4 Why language?

Whatever the genetics of language, a larger issue remains: Why would we remain focused on the autonomy of language – on "language . . . in and for itself," as Saussure (1959: 232) had it? In fact, a series of recent studies of primate comparison have emphasized the interactive precursors to language at the cost of genetics and an innate language faculty. Thus, Tomasello (2006) suggests that rather than placing human forms of communication at the center of evolutionary inquiries one should start with the foundation on which they rest, with uniquely human forms of collaborative engagement, including shared intentionality. "Many of the aspects of language that make it such a uniquely powerful form of human cognition and communication," he argues, "are already present in the humble act of pointing. And so in searching for the phylogenetic roots of human linguistic competence, we might profitably begin with the pointing gesture" (Tomasello 2006: 518). Apparently, apes do not point declaratively; they may point imperatively with an individualistic motive, to say "Gimme water!" or something like that, but they do not simply summon our attention by saying "Look!" to share an attitude about a referent. If this is the case, for chimpanzees gestures are procedures for getting things done, not a matter of intersubjective experience. One wonders, however, if honeybees point declaratively in the process of dancing. Or sperm whales with their sonar clicks.

Given Tomasello's perspective, "[a]sking why only humans use language is like asking why only humans build skyscrapers, when the fact is that only humans, among primates, build freestanding shelters at all" (2006: 520). Similarly, Levinson (2006: 44) maintains that through our fixation on innate linguistic structures we have overlooked the critical issue of everyday human interaction, the "interaction engine," "a set of cognitive abilities and behavioural dispositions that synergistically work together to endow human face-to-face interaction with special qualities" – a precursor, perhaps, to the virtual social networks of Facebook and Twitter. Many experimental studies have followed such leads, emphasizing chimpanzees' "theory of mind" and prosocial behavior such as generosity and altruism. Schmelz, Call, and Tomasello (2011: 2), for instance, conclude from their study of chimpanzee mentality: "If we define thinking as going beyond the information given in perception to make inferences, we may conclude that not only is thinking not the exclusive province of human beings, but thinking about thinking is not either."

If there is some kind of innate language device, organ, or faculty, it should be manifest in some universals common to all languages. Assuming that we adequately cover or reflect the variety of human language, it is pertinent to ask, as Foley (2005: 46) does: "If the knowledge of language is innate, simply the flowering of a pan-human 'language instinct,' how can we account for the obvious significant variation in the structure of human languages?" Some studies have introduced computational phylogenetic methods to address the nature of constraints on linguistic diversity in an evolutionary framework (Dunn *et al.* 2011). Interestingly, however, while geneticists claim to have located "the language gene" accounting for unique characteristics of human communication, linguists are busily challenging the core Chomskyan notion of language universals. There seem to be good grounds for taking any claims about language universals with a grain of salt. For one thing, claims about universals have been questioned on the grounds of a few detailed empirical studies of non-Western languages. Thus, both Foley (2005) and Levinson (2009) have objected to Pinker's (1994: 284) claim that "in all languages words for objects and people are nouns ... words for actions and changes of state are verbs." Foley (2005: 59) concludes on the basis of studies of Tongan and Tagalok: "Whatever is innate or 'instinctual' ... – and therefore fixed – in humans acquiring language, it can be neither the noun-verb distinction nor any connection between nouns, objecthood, and categorization of verbs, eventhood, and predication. These connections are learned, not preset, because ... the lexicon and grammatical patterns can be built otherwise."

There may be universals, but they are not of the kind normally assumed. Generalizations, however, are difficult if not meaningless due to the problem with variation; only a fraction of the 5,000 or more languages spoken today have been documented and studied by linguistic methods (Evans and Levinson 2009). The issue of language variation is not a trivial one. One of the persistent problems with generalizations about the structures of language is that they have been based on a limited and possibly skewed sample: "One could probably justifiably characterize the efforts of much scientific (and not so scientific) linguistics in [Greco-Latinate] ... languages over the last few centuries as a kind of linguistic imperialism (killing off indigenous languages being another, more pernicious kind)" (Foley 2005: 60). Ironically, keeping in mind Derrida's concerns with animals – with pluralizing the "animal," the brutality of humans to non-humans, and granting voice to other species – this kind of imperialism is echoed in Derrida's own work (2008), in the lack of sensibility to ethnographic variation.

7.5 Human–animal relations of production

Non-human animals play an increasing role in the biotech mode of production, mainly in the context of biomedical experiments and the production of organs for human use. How should one typify human–animal relations in this context? While Marx did address the work carried out by animals at several points, some of his characterizations, as already suggested, are likely to sound anthropocentric, narrow, and outdated to many modern readers. Reluctant to allow for the possibility that animals might "produce," he was at pains to make a radical distinction between humans and animals in this respect:

> In creating a *world of objects* by his practical activity, in his *work upon* inorganic nature, man proves himself a conscious species-being ... Admittedly animals also produce. They build themselves nests, dwellings, like the bees, beavers, ants, etc. But an animal only produces what it immediately needs for itself or its young. It produces one-sidedly, whilst man produces universally ... An animal produces only itself, whilst man reproduces the whole of nature. (Marx 1959: 276; emphasis in the original)

Drawing upon Marx's critique and at the same time going beyond his anthropocentrism, Haraway's work *When Species Meet* is a major treatise on many of the issues invited by the biomedical mode of production – the equivalent, perhaps, in the domain of human–animal relations to Marx's *Capital, vol. 1* (appropriately, early on Haraway seems to have used the working title of "Biocapital, vol. 1" for her work). Haraway (2008: 62) emphasizes that humans and their "companion species" emerge as "mutually adapted partners in the nature-cultures of lively capital" and that it is time to think hard about the "encounter value" generated by such mutual adaptation. This encounter value, she suggests, remains underanalyzed and might be more fruitfully addressed by returning to Marx than by attending to the now fashionable bioethics of animal rights: "The Marx in my soul keeps making me return to the category of labor . . . My suspicion is that we might nurture responsibility with and for other animals better by plumbing the category of labor more than the category of rights, with its inevitable preoccupation with similarity, analogy, calculation, and honorary membership in the expanded abstraction of the Human" (Haraway 2008: 73).

There is no good reason, indeed, for excluding the generative powers of animals from the production process. Many foragers present their prey as benefactors engaging in mutual collaboration with humans, in a somewhat Marxian fashion: "Marx could hardly have imagined an Algonquian labor process in which humans and animals successively participate as producers of the other, the animals willingly surrendering the 'product' of their own bodies and the hunters returning it to them as cooked food, all figured in the idiom of 'love.' But his reflections on an authentically social labour process are evocative of the benefactive model of Cree–animal relationships" (Brightman 1993: 188).

With advances in functional genomics, cross-species research has reached new levels, manufacturing animals that serve as substitutes for human experimentation, testing the limits of anthropocentrism, and posing engaging questions about liminality and ethics. As Hoeyer and Koch (2006: 387) point out, cross-species research in genetics "erodes" the distinction between humanity and animality, challenging the notion of unique human worth, one of the fundamental notions of bioethics frequently introduced in debates on research and experiments involving human embryos. Here, as elsewhere, it may be essential to relax or destabilize some of the ethnocentric assumptions of the grand narratives of Western biology and bioethics. Drawing upon her ethnography from a Papua New Guinean context, Bamford (2007: 27) argues that "[i]f 'crossing' species boundaries is threatening to the social order of the West, for Kamea, by contrast, it is constitutive of it."

7.6 Conclusions

It seems likely, given the evidence of recent primate studies, that human language is just the tip of the evolutionary iceberg, based on the broad foundation of sociality that we share to one degree or another with at least our closest relatives. How to read this theoretically, however, is another issue. One common reading is the dualistic one, assuming the separate but interlinked systems of biology and culture, nature and society (Richerson and Boyd 2008). A rather different reading, drawing upon developmental-systems theory and related developments in several fields – in particular, anthropology, psychology, and philosophy (Descola and Palsson 1996, Oyama 2000, Ingold 2001) – seeks to collapse biology and culture. The great challenge is to picture humans, chimpanzees, and other beings as constituted by, and embedded in, a single, integrated ensemble of biosocial relations (Palsson 2013), and to explore what such a perspective might entail for the understanding of communication, language, and cognition as they unfold in the stream of life and, more broadly, for the understanding of the human condition and for the refashioning of disciplines traditionally erected around the nature/society divide. The "nature" with which we are born and which we develop is thoroughly biosocial, embodied through human activities.

Cognitive nativists and evolutionary psychologists (see, for instance, Pinker 1994) draw upon Chomsky's notion of the language faculty to argue that human language is unique, a pan-human skill engraved in our genome thanks to selective pressures during the hunter-gatherer past, pointing out that it is effortlessly acquired by children and universally manifested in the diverse languages of human history and dispersal. Judging from the accounts just discussed, the language universals are still on the run and nativist theory about the uniqueness of human language remains unconvincing. Interestingly, in his co-authored *Science* article Chomsky seems to have distanced himself from the thesis of human uniqueness: "The available data suggests a much stronger continuity between animals and humans with respect to speech than previously believed. We argue that the continuity hypothesis thus deserves the status of a null hypothesis, which must be rejected by comparative work before any claims of uniqueness can be validated. For now, this null hypothesis of no truly novel traits in the speech domain appears to stand" (Hauser, Chomsky, and Fitch 2002: 1574). For some scholars, this must count as a radical shift in perspective, given Chomsky's earlier writings, a shift that seems to have irritated some of his followers (see Haraway 2008: 235). Yet, although the conclusion of the *Science* article may have violated some linguistic orthodoxy (Anderson 2004), for Chomsky it need not have represented a major move. In his *Knowledge of Language* he suggested that the language faculty is "a distinct system of the mind/brain, with an initial state S_0 common to the species ... and *apparently* unique to it in essential respects" (Chomsky 1986: 25; emphasis added), adding, however, an important footnote: The questions of "innateness and species-specificity are distinct" (1986: 48, fn. 13), he argues, emphasizing that he has avoided confusing the two. Perhaps, then, *both* chimpanzees and humans acquired a language device in the course of evolution, and possibly sperm whales and honeybees as well. Alternatively, it might be argued, *neither* the language of humans nor that of chimpanzees and other "lower" creatures is best attributed to anything that may be identified as an innate language device. Perhaps the *absence* of universals, too, should be taken as the null hypothesis, to be challenged by further comparative work.

A few decades ago, anthropologists imagined they could safely cling to the cliché that only humans were capable of tool making, inferences, subjectivity, and, above all, language – in sum, of making and having "culture." In spite of, perhaps because of, their focusing on comparison and evolution, their perspective has largely remained anthropocentric. Step by step, however, each of the traditional indicators presumed to demarcate human culture and mentality from the natural world of non-human animals – tool use, signing, cultural transmission, emotional bonding, sociality – have proved to be erroneous, flawed, exaggerated, or far more complex and tricky than previously anticipated. And language no longer occupies center stage. Despite all the attention he received, Nim Chimpsky may not have made history, at least not in the context of theory of mind and language. However, while he didn't contribute much to syntax and semiotics, he was clearly capable of responding, playing, and suffering. If ever there was a serious attempt to cultivate a chimpanzee (or for that matter, any non-human animal) – to "cast off apehood," as the story of Red Peter had it – it was Project Nim.

As we have seen, chimpanzees are often presented as a border species, almost endowed with speech, almost human, almost in culture. Reaching out to chimpanzees, whether through sign language in experimental settings or participant observation in the wild, has often been rhetorically represented as communication across the nature/society divide, as a handshake through millions of years. The researchers involved seem to imagine they have managed to gallop back through evolution – much as Neil Armstrong triumphantly announced a "giant leap for mankind" across space, when landing on the moon. As a liminal species, chimpanzees keep attracting our attention, offering a kind of view-from-afar on us, across great divides. Such divides, Fox Keller (2000) suggests, represent a persistent, if not unavoidable, "mirage." For a growing number of scholars in a variety of disciplines it seems both essential and feasible to move beyond the mirage. In the absence of a better non-dualistic language, the notions of becomings and biosocial relations (Ingold and Palsson 2013) may help to challenge current understandings of the division of biological and social anthropology and their essentialist takes on key issues, including those of human nature and relatedness and the interdependencies of humans and other kinds of beings. Chimpanzees, just like humans, are ensembles of biosocial relations, relational beings continually becoming with other beings, including humans.

8 Lucy in the sky: celestial bodies

Ever since humans ventured beyond their cradle in East Africa, they have advanced their exploration of the rest of Earth step by step, with the most spectacular successes during the past two centuries. Now that most of the habitat of the globe has been charted and conquered, humans are increasingly turning their attention to the "remotest corners" of both living organisms and outer space. Focusing on languages of voyaging and mapping, this chapter discusses the similarities and differences between research on what biologists sometimes refer to as the "universe" within the human body (in particular the foetus and the genome), on the one hand, and, on the other hand, the exploration of outer space. The enterprises of the Big Sciences of molecular biology and astrophysics, I argue, zooming in and out, are related in many ways – and not just through similar languages and metaphors.

During the past half century, both molecular biology and space science have made immense progress. Some of the significant events generated by the former were Rosalind Franklin's photography of DNA (1952), the discovery of the double helix (1953), and the drafting of the human genome (2000). Among the critical events taking place in space during the second half of the twentieth century were the launching of Sputnik (1957), blasting Gagarin into orbit (1961), landing humans on the moon (1969), and establishing the International Space Station (1998). To celebrate half a century of human exploration in space, the *Guardian* (2007) published a series of photographs showing astrophysicist Stephen Hawking, accompanied by his assistants, floating on a zero gravity jet above the Florida coastline. The plane made several parabolic dips, giving Hawking the experience of weightlessness. The images powerfully underline both the mastery of space and the dependence and fragility of the human being. Newton's apple is not far. Or is it perhaps a reference to the sinfulness of human digression into the heavens, a modern version of the story about the Tower of Babel?

Since the European Middle Ages, the notion of "celestial bodies" or "celestials" (from *caelum*, the sky) has usually referred to heavenly bodies belonging to reigning gods or emperors on Earth. With the advances of molecular science, including those of genetic engineering and synthetic biology, the idea of constructing corporeal bodies *for* outer space has become possible. While hitherto human exploration has taken place without subspeciation, this is unlikely to be the case in the event of human settlement in outer space. Indeed, the future may bring forth a post-human settler in space, a celestial "Lucy in the sky" – or, more likely, several kinds of them.

8.1 Foetal space

Several scholars have drawn attention to the parallels between imagery relating to, on the one hand, human foetuses and the womb, and, on the other hand, the celestial world. As Michaels (1999: 125) points out, "[p]lanets, supernovas, and galaxies have been showing up alongside foetuses, embryos, and blastocysts during the past twenty-five years, and their visualization occasions comparable journalistic indulgences and epistemic quandaries." Duden (1993) explores the path-breaking photographs of Lennart Nilsson documenting "the beginning of life," published in the *Life* magazine in 1965 and 1990, pointing out that the photographs, the first ones of their kind, and the accompanying text in *Life* repeatedly juxtapose foetuses and astronauts, bodies and space. The 1990 issue elaborates on the first stages of human life by means of celestial language and imagery, starting with a foetus aged two hours:

> Like an eerie planet floating through space, a woman's egg or ovum ... has been ejected by one of her ovaries into a fallopian tube ... The luminous halo around the ovum is a cluster of nutrient cells feeding the hungry egg. (Quoted in Duden 1993: 13)

Eight days later, *Life* goes on:

> The blastocyst has landed! Like a lunar module, the embryo facilitates its landing on the uterus with leg-like structures composed of sugar molecules on the surface. (Quoted in Duden 1993: 14)

The fascination with the foetus and outer space underlines human curiosity at the border of the unknown, the urge to extend one's sight by zooming in or out beyond the "natural" horizon of the human eye. New images and visual horizons, however, sometimes take on a life of their own, informing perception and politics. Thus, astronauts tend to speak of a gestalt shift as a result of their voyages into space, when seeing Earth from a distance, a "Gaia" perspective that seems to facilitate global, environmental concerns. No doubt, images from space, including images from Neil Armstrong's "giant leap for mankind" when landing on the moon, have also informed public discussions on Earth on several scores. Likewise, the kind of fatal imagery presented by the famous *Life* photographs just discussed is likely to have had an enormous impact on public discussions of biopolitics – in particular, abortion and the rights of women. Duden (1993) addresses the issue with reference to what she calls the

"Nilsson Effect," highlighting the shift in emphasis as a consequence of Nilsson's photography from the pregnant woman to the foetus and the resultant alteration in power balance between those advocating "pro choice" and others in favor of "pro life."

Some other works make similar analogies and connections between foetuses and astronauts, from the vantage point of outer space. An article by Sofia (1984), largely devoted to a commentary on the film *2001: A Space Odyssey* (1968) by S. Kubrick and A.C. Clarke, suggests the film establishes what she calls "Jupiter Space" through the imagery of the foetus, a space "whose contours are elaborated in visual complexes which equate the male brain, the womb, outer space, city landscapes, grids of light, microcircuits, the interiors of computers, skyscraper façades, and so on" (1984: 48). More generally, she argues, the human foetus operates as a symbol for Earth: "It is a cosmic symbol. It is not entirely inappropriate that the planet be represented by a signifier of unborn life, for it presently contains all of the possibilities for future life forms. From this perspective, disarmament might be seen as an act to prevent a cosmic abortion" (Sofia 1984: 56). While Sofia's article is highly playful and speculative, it draws attention to military concerns and the Cold War and their impact for both biopolitics and space exploration. With the development of digital technology, bioinformatics, and the new genetics, voyaging into the human body was escalated and extended to specific organs, in particular the brain, cellular material, and the genome.

Eventually, "Man the Hunter" and "Woman the Gatherer," made in East Africa a long time ago, turned to hunting and gathering in their own genome. What project could be more anthropological? *Homo viator* navigating waters right at home. One of the characteristics of the genome era is the application of a cartographic language to the human body. In year 2000, the first draft of a "map" of the human genome was announced. On that occasion, the journal *Nature* triumphantly invited its readers on a tour into the "universe within" with the following grand statement:

> Since ancient times we have drawn charts of the sky, of the world, and of our anatomy. Today, a new chart is added to the collection: The map of our genome. Its purpose is to synthesize the insights and meaning gained from the sequence of the human genome. We invite you on a tour of the geography of the genome, exploring the chromosomes, the sequence, and the differences between individuals and populations. The integration of these exciting new findings ushers in a new era of scientific and medical progress. (*Nature* 2000: 395)

Impressed with their ability to zoom in on the minute details and contours of hereditary material and their power of visualization, geneticists and molecular biologists have firmly reinforced their language of cartographies, a language that echoes the modernist notion of expansion and mastery.

8.2 Out of Africa, out of Earth

In 1974, a 40 percent complete *Australopithecus afarensis* skeleton was discovered in the Awash Valley of Ethiopia's Afar Depression. The fossil (AL 288–1) was nicknamed Lucy after the Beatles song "Lucy in the Sky with Diamonds," which was being played repeatedly at the time of discovery in the archaeological camp at Awash Valley. Lucy was estimated to have lived 3.2 million years ago, firmly establishing human origins in East Africa. For a long time there has been a rich discussion of the roads out of Africa, their timing and significance for the understanding of hominid evolution and history. Recent archaeological findings from Dmanisi in the Republic of Georgia shed new light on some of the issues involved, possibly documenting the earliest members of the genus *Homo* outside the continent, thereby filling in significant gaps in current knowledge about a critical phase in human evolution (Lieberman 2007). Dated to 1.77 million years ago, the fossils involved indicate great variability in body and brain size, in the size range of *H. habilis* and *H. erectus*, reflecting, among other things, selection for improved terrestrial locomotion. These findings, then, provide an image of scattered and variable hominids adapting to time and place outside Africa. Will there be a parallel development in outer space, in the event of human settlement on different asteroids, space stations, moons, and planets?

One of the most important contributions to the anthropology of space is the volume *Interstellar Migration and the Human Experience* edited by Finney and Jones (1985). Written in the heat of the Cold War, under the threat of nuclear war, it represents a particular "galaxy of discourse," to borrow a term from another important contribution to the field (Battaglia 2005). For Finney and Jones (1985: 5), it is human "biocultural" nature to venture into new areas, to explore the entire globe and eventually head for the sky: "Sometime in the not too distant future a space traveller will do something science fiction writers have been talking about for decades: A human being will jump completely off a small world. Asteroids and small moons have very weak gravity indeed." "Barring total nuclear war, a devastating collision with a comet or asteroid, or some other calamity on a worldwide scale," they conclude, "there is a good chance that this initiative will soon result in settlement in near space and that eventually our descendants will scatter among the stars" (Finney and Jones 1985: 333). No doubt, the birth of the first human child in outer space will be an event of enormous symbolic significance, underlining human *settlement* outside Earth in contrast to the relatively brief visits of the past, much like the birth of the first child outside Africa.

The likely consequence of human settlement in space is one of the themes explored by Finney and Jones (1985). For a long time, they suggest, the hominid species has formed one interbreeding world population and speciation has not seemed possible. That, however, would not hold in space since our descendants will probably be scattered throughout the vastness of space that would set up the conditions for the rapid speciation of *Homo sapiens*:

> If the technology of space colonization really works, if our descendants do settle throughout the Solar System and then migrate to other star systems, humanity will never be the same again. The course of human evolution will change utterly and inalterably ... The threshold of space is also the threshold to quantum biological evolution. (Finney and Jones 1985: 22–23)

Time, no doubt, has complemented and qualified the predictions of Finney and Jones on several scores. For one thing, the obvious post-Cold War candidate for what they refer to as "some other calamity on a worldwide scale" is not nuclear war but massive environmental change generated by humans, including global warming. Our successes at exploration, colonization, and resource use have made Earth a rather messy place for humans and many other species, and unless the problems at home become paralyzing this is likely to escalate the exploration of outer space. Colonization and modernization have been taxing to the planet, a point emphasized by Mahatma Gandhi. Asked whether independent India would follow the British pattern of development, Gandhi replied: "It took Britain half the resources of the planet to achieve this prosperity. How many planets will a country like India require?" (quoted in Moran 2006: 150). How many planets will the world require to satisfy the needs of the masses and to clean up the mess?

Finney and Jones (1985) did not anticipate (or at least they did not comment upon) the successes of the new genetics and their potential relevance for settlement in space. Thanks to the spectacular advances of biotechnology and genetic engineering long-term biological adaptation to outer space is no longer pure science fiction. As Rheinberger (2000a: 19) suggests, modern gene technology along with the molecular biology developed between 1940 and 1970 facilitate "the prospects of an intracellular representation of extracellular projects – the potential of 'rewriting' life." The key tools of recombinant DNA work are not "sophisticated analytical and electronic machinery" but "macromolecules that work and perform in the wet environment of the cell ... The scissors and needles by which the genetic information gets tailored and spliced are enzymes. The carriers by which it is transported into the cells are nucleic acid macromolecules" (Rheinberger 2000a: 24–25). As a result, Rheinberger argues, the traditional dichotomy between "nature" and "culture" no longer makes much sense. Quite possibly, in the future life will be "written" and "edited" for outer space, inviting new kinds of citizenship and biosocialities and new kinds of hybrids of technologies and organisms. Some of the early speculations along these lines are those of Sofia (1984). If the Earth is an embryo, Sofia suggests,

> then its womb is space. Although we know of no other living worlds, centuries of extraterrestrial fantasies capped by several decades of off-world practice have encouraged us to think of space as a good womb, full of inhabited planets. From this view, the Earth is just one of many cosmic pregnancies. It doesn't really matter if we abort it, for we can always escape to one of the new Star Children we pluck from the vacuum; we might even mutate into extraterrestrial cyborgs. (Sofia 1984: 57)

Not only does modern biotechnology increase the likelihood of human and post-human variation, encounters with radical "others" from space cannot be ruled out. Ufology, as Roth (2005: 39) points out, the study of unidentified flying objects and extraterrestrial visitors, "is a discipline that has tried to understand racial diversity. Ufologists do not always call it 'human' diversity, but then the earliest European anthropologists were not sure that all speaking bipeds outside Europe were human either." The difficulties of adapting to permanent settlement in space should not be underestimated, in particular the damaging effects of radiation on human bodies and problems relating to what might be called the "phenomenology of space," the challenges posed by our Earth-bound perceptual, cognitive, and psychological capacities in the context of space. Addressing the relations between humans and the material world from a phenomenological perspective, Ingold suggests what he calls "a view from the open"; rather than imagine "that life is played out upon the surface of a world already furnished with objects," he argues, people "make their way *through* a world-in-formation rather than *across* its pre-formed surface. For that reason, the fluxes of the medium through which they move are all-important" (2007: S32; emphasis in the original).

Such a view emphasizes the everyday human experience of wind and weather. There is no compelling reason, however, to imagine that the "view from the open" does not apply to life in space; in fact, it may be even *more* relevant for astronauts than earthlings. Space is a particular kind of medium with its own formations and fluxes – a medium where everything floats, where solar particles ("winds") blur vision, and where Earth may not even be in sight. While space, much like the alien medium of water, poses particular problems for a species thoroughly adapted to terrestrial life, these can at least partly be suppressed by means of technological and informatic enhancements like robots and artificial intelligence. The mere fact that settlements in outer space *are* a possibility underlines that space is just as "natural" for humans as Earth.

There are profound problems to face in outer space, and, indeed, they may slow down attempts to establish human settlements outside Earth. Some of the problems likely to be encountered are unprecedented biopolitical and bioethical issues that will inevitably be distracting, issues relating to inequality, difference, citizenship, race, and, possibly, eugenics. Humans, however, have been moving fast on the space front for the past half a century and their journeys seem likely to continue. Given the spectacular advances in travels in outer space over the past decades, following the launch of Sputnik, it is quite conceivable that in the future life will be "edited" for outer space, inviting new kinds of citizenship and new kinds of hybrids of technologies and organisms, new kinds of celestial bodies. While such issues have long been relegated to fiction (for recent examples, see Moseley 2002 and Crichton 2006), now they are very much on the academic agenda (Valentine, Olson, and Battaglia 2012). The original meaning of the title of "Lucy in the Sky with Diamonds" has always been a matter of debate, but whatever John Lennon's reasoning when writing the lyrics – in one theory there was a psychedelic element induced by drugs (LSD) – Lucy seems to be heading for the sky.

III

Biospheres

In this section attention moves to situated knowledge, human–environmental relations, the importance of ethnography for environmental discussions and policy, attempts to move beyond the nature–society dualism, and the challenges of the Anthropocene. The broad environmental context within which we are embedded is unique in the sense that more than ever before the environment is the *result* of human activities and, moreover, humans are acutely aware of this. Since our environment is increasingly refashioned by ourselves, it is suggested, we need to radically reframe global environmental problems as fundamentally human – moving beyond the traditional, narrow confinement in both academic and public discussions of the "environment" to the "natural" domain. The last chapter in the section explores the usefulness of the notion of housekeeping, of house models and the *Oikos*, for contemporary environmental politics and debates. The current collapsing of body, society, and environment, in both biosocial theory and the "real" world, necessarily invites applying the governmental gaze throughout, from the cellular to the global level, embedding humans and other beings in ever-larger contexts across the different scales of the Anthropocene.

9 Enskilment at sea: situated knowledge

To take a decentered view of master–apprentice relations leads to an understanding that mastery resides not in the master but in the organization of the community of practice of which the master is part . . .

Jean Lave and Etienne Wenger, 1991[1]

When talking about seasickness, Icelanders implicitly recognize the relationship between knowledge and practice, and the unity of emotion and cognition, body and mind. For them, "seasickness" (*sjóveiki*) not only recalls the bodily state of nausea sometimes caused by lack of practical knowledge, the unexpected rocking movements of the world, but it is also used as a metaphor for learning in the company of others, which is seen in terms of the *recovery* from seasickness – "getting one's sea legs" (*sjóast*). Those who return "home" after an experience on rough seas are said to have *found* their sea legs (*verða sjóaðir*); as nausea is replaced by well-being, a quantum leap in learning and sociality takes place. I shall argue that getting one's sea legs – becoming skillful – means to attend to the task at hand, actively engaged with a social and natural environment. This suggests a notion of enskilment that emphasizes immersion in the practical world, being caught up in the incessant flow of everyday life – and not simply, as many cognitive studies have assumed, the mechanistic internalization and application of a mental script, a stock of knowledge or a "cultural model," "what one needs to know in order to behave as a functioning member of one's society" (Holland and Quinn 1987: 4).

The approach adopted here is informed by theories of practice and practical knowledge recently advanced in several disciplines, including anthropology (Bourdieu 1990, Lave 1988, Ingold 1993). Such an approach was outlined many years ago in both European and North American contexts – by Malinowski (see Leach 1970), Polanyi (1958), and some others (Joas 1993); only recently, however, has it begun to emerge as a consistent theme with a momentum of its own (Ortner 1984), reinforced by current theorizing on

[1] J. Lave and E. Wenger, *Situated Learning: Legitimate Peripheral Participation*, 1991, Cambridge University Press.

the body. The perspective of practice theory, I suggest, has important implications for studies of differential fishing success, a theme explored by a number of anthropologists, especially discussions of the so-called "skipper effect," the idea that the contribution of the skipper is critical to the size of his catch relative to that of others (Palsson and Durrenberger 1982, Gatewood 1983). Discussions of differential fishing success, I argue, often remain couched in reductionist terms, emphasizing the notion of the autonomous individual, sometimes at the expense of both ethnography and theoretical insight. Acheson observes that the "technical skills" of the fishing captain "are far more critical in determining the catch of a boat than *any other single factor*" (1977: 111; emphasis added) and much research has attempted to establish to what extent that is so. The problem, however, is that it is far from clear what exactly the reference to "technical skills" as a "single factor" should be taken to mean.

Readdressing the issue of fishing skills, and emphasizing the social nature of human action, inevitably shifts the ethnography to new kinds of questions. Focusing on everyday activity and situated practice rather than individualistic, normative models – as Lave (1988: 15) points out in another context – "motivates ... a different set of problems and questions than the study of virtuoso performance and people's failures to produce such performances." In particular, it encourages us to attend to whole persons, master–apprentice relations, and the wider community to which they belong, decentring the study of enskilment. If enskilment is a necessarily collective enterprise – involving whole persons, social relations, and communities of practice – so also is ethnographic apprenticeship, doing anthropological fieldwork (Gudeman and Rivera 1990, 1995). I will argue, drawing upon my experience in my native Iceland, that enskilment in fieldwork inevitably involves psycho-somatic processes, if not veritable "gut reactions." Fieldworkers usually begin their "trip" on the margin of the community, nauseated by their novice status. As they become increasingly involved in and knowledgeable about the activities of others, they move towards the center and begin to feel "at home," in both their bodies and the company of others.

9.1 Learning theory

Theories of learning and craftsmanship have often been presented in highly normative terms, presupposing a natural novice who gradually becomes a member of society by assimilating its superorganic heritage (see Bloch 1991: 189). Although different social theories have different notions of the relationship between the individual and the collective and of what sociality entails – thus Boasian culture theory speaks of the "enculturation" of individuals as cultural knowledge is stored in their heads, while for Durkheimians "collective representations" are supra-individual byproducts of interaction – in *normative* theory learning is generally assumed to take place with socialization, broadly defined as the acquisition of a stock of knowledge about expected ways of thinking, feeling, and acting (see Ochs 1988: 5). Given the normative approach, learning entails the transmission of culture, a mental code or script that exists prior to and independent of human activities, a recipe for action (a pre*scription*) analogous to a book of grammar or a dictionary – in short, a Saussurean *langue*. Related to the notion of learning as cultural transmission is the tendency to think of the person in terms of a container, "as if the person possessed a fixed capacity [of competence] analogous to the amount of liquid that can be placed in a glass" (Fischer *et al.* 1993: 94).

There are several serious problems with such an approach. For one thing, in reducing the novice to an imitator of technique, the normative approach leaves unanswered the fundamental question as to how a body of cultural knowledge is constructed. If both novices and tutors (by definition, former novices) are mere recipients of models and texts, submitted to a body of knowledge given in advance, it is difficult to see how human creativity could be possible at all. This problem is evident in much of the literature on "indigenous knowledge" and "ethnoecology"; as Ellen (1982: 225) remarks with respect to this literature, people "just do not conduct their lives in . . . pre-programmed ways." Also, the model of cultural transmission is highly ethnocentric – the product of Western history and textual discourse, reinforced by the tradition of literacy and the institutions of formal schooling and disembedded training – caught up in dualisms of mind versus body and learning versus context (Suchman 1987). Western tradition is preoccupied with analytic and theoretical ways of knowing, *episteme*, devaluing and misrepresenting practical and contextual knowledge, *techne* (Marglin 1990). This tradition needs to be explained and not taken for granted in a truly comparative study of epistemology.

The normative view, moreover, misconstrues the essence of the lived-in world, failing to capture what it means to engage in a skillful act, the "feel" for the game. Rosaldo (1980: 22) rightly remarks, using the game of basketball as an example, that ethnographic narratives of complex events must go beyond descriptions of rules and strategies; "there is a vast difference," he points out, "between merely knowing the rules of the game and having the knowledge required to follow the game in the manner of an experienced fan of fine judgement." What is most important, however, is the construction of a third kind of knowledge, conveying the essence of the game *from the point of view of the player*. This is precisely the approach Rosaldo seems later to have adopted, in his critique of the objectivist ethnographer who "positions himself as a spectator who looks on from the outside" (1987: 99). While rules and representations are relevant for the participant – the player – they are not what learning is all about. As Ingold (1993: 462) argues, the novice becomes skilled not through the acquisition of rules and representations, "but at the point where he or she is able to dispense with them":

> rules and representations ... are like a map of an unfamiliar territory, which can be discarded once you have learned to attend to features of the landscape, and can place yourself in relation to them. The map can be a help in the beginning to know the country, but the aim is to learn the country, not the map.

Indeed, a person who cannot *use* his or her knowledge to do something is "like a man who collects maps but never takes a trip" (Miller, Galanter, and Pribram 1960: 2).

Finally, normative theory assumes a one-way, hierarchical ordering of knowledge, as Lave (1988: 8) points out:

> In this theory, duality of the person translates into a division of (intellectual) labor between academics and "the rest" that puts primitive, lower class, (school) children's, female, and everyday thought in a single structural position *vis-à-vis* rational scientific thought.

Normative theory, then, necessarily dismisses everyday language as cheap and theoretically irrelevant, as "mere" household words or "loose talk."

Oddly enough, anthropologists, who at the beginning of fieldwork normally assume the role of the apprentice, sometimes shift to the heroic role of the privileged "observer" at a later stage in their work; when in the field, in the process of writing *down*, they present themselves as naïve novices and their "informants" as their tutors, but as soon as they return to their academic bases the roles are sometimes mysteriously reversed, as if the act of writing *up* necessarily presupposes a kind of communicative metamorphosis. The humble novice anthropologist becomes a tutor. The textual "ups" and "downs" of ethnographic research are separated not only by physical movement; the journey "back home" represents a rite of passage during which the anthropologist supposedly experiences an instant coming of age – and meanwhile the native is trapped in the mythical "ethnographic present." Malinowski ridiculed the validity of native accounts – even though they might turn out to be close to or identical to his own. Malinowski, Leach (1970: 134–135) suggests, "specifically mocked at the account of Trobriand social structure that one might expect to obtain from a professional Trobriand informant, ... though when he himself attempted to write a concise description of 'The Constitution of Trobriand Society' ... the result resembles most strikingly that given by his imaginary despised 'informant.'"

Practice theory offers an alternative view of learning, craftsmanship, and ethnography – a view that allows for a novice whose sociality is given right from the beginning. Assuming a social or constitutive model of the individual is to introduce purpose, agency, and dialogue into the process of enskilment – a radical break with the Cartesian tradition of separating ideas and the real world, learning and doing, experts and laypersons, knowledge and practice. The theory of practice, then, suggests a change in gestalt which "shifts the boundaries of activity well outside the skull and beyond the hypothetical economic actor, to persons engaged with the world for a variety of 'reasons'" (Lave 1988: 17–18); the proper unit of analysis is no longer the autonomous individual separated from the social world by the surface of the body, a natural being who passively internalizes the mental scripts of the cultural environment, but rather the whole person in action, acting within the contexts of that activity.

Such an approach – informed by practice theory and the notions of situated action, mutual enskilment, and communities of practice – emphasizes both democratic communion and the continuity of the social world. While the relations of novices and their tutors are rarely those of equals, there is quite a difference between the open and linear system of apprenticeship and the dualistic structure suggested by normative learning theory. The former mode, as Reed (1993: 54) remarks, invoking the "ecological" language of J.J. Gibson (1979), involves joint intentions and collaborative instruction, "in the sense of helping others to learn the affordances of objects and tools within the context of a given skill," whereas the latter involves a pyramidal context and a formal method of teaching.

9.2 "Getting one's sea legs"

Moving from learning theory to ethnographic practice, my own Icelandic fieldwork provides a useful starting point. In the early phases of my fieldwork, the process of establishing the necessary trust and rapport often seemed painfully slow. At one point, however, in the middle of the winter season in 1981, I felt as if I had made certain and sudden progress. A skipper invited me to join him on one of his fishing trips: "If you *really* want to know what the fishing industry is all about, you must go fishing." I accepted his offer, and shortly after we left harbor we were in rough seas. While the skipper and his crew patiently waited for the weather to improve, however, I was busy emptying the contents of my stomach. Then, miraculously, it seemed, the weather improved and, fortunately, the feeling of seasickness vanished. Quite suddenly, my nausea had been replaced with a sensation of alertness and well-being, an "oceanic" feeling. I had, quite literally, "found my sea legs."

Many other fieldworkers have referred to similar "breakthroughs" – signals of dramatic, positive change – in their relationships with the people they visit (Okely 1992: 17); such stories are often used in modern fieldwork rhetoric. For me, however, the metaphoric association with the journey draws attention not so much to privileged excursions into the exotic as to the similarities in doing fieldwork and fishing in rough seas. Seasickness, in fact, has much in common with the psycho-somatic symptoms identified, by both fieldworkers and medical experts, as "culture shock" (see Wengle 1988: 9–10). It would be wrong, of course, to simply dismiss the experiences of seasickness and culture shock as pathological ones; both involve bodily manifestations of a particular stage in learning. The fishing trip, then, served as a useful reminder of a critical stage in my fieldwork, and, more generally, of the emotional and physical manifestations of mastery and enskilment. As many had before me, I came to realize that to engage in anthropology is not just to "observe" and record but to participate in the lives of other people.

While I find it useful to refer to my fieldwork as a fishing trip, I do not regard it as a solitary trip across cultural boundaries. During my original fieldwork, of course, I relied heavily on conversations with local friends and key informants, and later on I collaborated extensively with another anthropologist (Paul Durrenberger). Whether we liked it or not, we were rarely able to enjoy the luxury of being the detached ethnographer. Anything we had to say on the subject of common interest – differential fishing success – inevitably got entangled in ongoing debates among indigenous fishermen and boat-owners on the role and nature of individual contributions, on fishing expertise and resource management. We came to the conclusion that anthropology was best described as a dialogic enterprise involving both feedback and disagreement, as a Malinowskian "long conversation" (Gudeman and Rivera 1990) in which the acts of "writing down" and "writing up" were difficult to separate. This recognition – triggered by discussions with fishermen and social scientists and by the reading of anthropological texts – sent me back to my early fieldnotes, to the communities where I had worked, and to comparative ethnography. Somewhat ironically, it was only some years after my experience of seasickness, fishing, and collaborative ethnography that I began to suspect that the perspective I had adopted with respect to my own enskilment as fieldworker and ethnographer might be applied to the context I had studied – to fish-finding and travels at sea.

In fact, the contrast made above, between the textual approach of normative theory and the situated approach of practice theory, may well be illustrated with reference to the ethnography of fishing and ocean navigation. An article by Gladwin (1964), which compares the strategies used by European navigators and by Trukese sailors when traveling from one island to another over miles of empty ocean, contains some pertinent observations. In the case of Europeans, Gladwin suggests, most of the necessary thinking is typically done in advance of its implementation, whereas in the case of the Trukese, decision-making is continuous, accommodated to winds, tides, ocean currents, and so on. Also, while European navigators are able to describe in words the procedures they follow and are likely to give a "logical" explanation of what they are doing, a Trukese navigator "cannot possibly put into words all of the myriad perceptions which have led him to be sure at that moment where the island is"

(Gladwin 1964: 174). The European navigator operates deductively, proceeding from general principles to details, whereas the Trukese navigator seems to operate inductively "[h]e does indeed start with details, but he never arrives at any discernible principles" (Gladwin 1964: 175). Gladwin's description, then, captures some of the differences between the abstract, textual logics of much of European navigation and the situated, intuitive procedures of Trukese navigators (see Hutchins 1983, Suchman 1987: vii–x). His analysis, however, tends to be phrased in the normative terms of Western educational discourse, emphasizing mental models and "cognitive strategies" – what goes on inside the brain: "The Trukese navigator," he argues, "does it all in his head. This is an astounding intellectual achievement" (Gladwin 1964: 172). Gladwin is quite right in pointing out that Trukese navigation is a remarkable achievement – we may refer to "this kind of ability as a 'knack,'" he says, "and respect a person for his competence" (1964: 175). Anxious to show, however, that Trukese are able to travel long distances on open boats without the use of the "external" navigational devices of the European, he accepts uncritically the mentalistic premises of the normative theory of craftsmanship.

In an article on salmon fishing in Alaska, Gatewood (1983) similarly emphasizes intuitive, situated procedures, contrasting them to "rational" procedures of decision-making by means of terms not unlike those used by Gladwin. For Gatewood (1983: 348–349), a "rational" procedure involves conscious consideration of alternatives, deliberation in terms of these considerations, thoughtful reflection on the process itself, and adherence to well-defined procedures for selecting the right alternative. An intuitive or "reasonable" procedure, he argues, is the same as a rational one "except that the manner by which the final synthesis is accomplished cannot be specified by the decider in advance of the decision itself" (Gatewood 1983: 348–349). In Alaskan fishing, the crew expect their skipper to make a decision on a "rational" basis – "as opposed to basing it on hunches, sheer randomness, dreams, divination, etc.," but contrary to the image of rationality they try to protect, under severe constraints of competition and impression management, "most skippers fall short of explicit, algorithmic procedures when it comes time to decide" (Gatewood 1983: 348, 349). Gatewood's explanation for the discrepancy between image and reality is not simply that rational solutions to the decision dilemmas of fishing are unavailable; more importantly, skippers' decisions are "a matter of unspecifiable skill, of tacit knowledge" (1983: 363). Shifting the focus of inquiry "from normative and outcome-predictive to ethnographic," Gatewood (1983: 364) argues, "reveals how people differ from the machines and mathematical models they create." Bloch (1991: 190) similarly points out that if we think of decision-making as a normative serial process carried on by a single processor, everyday tasks which a skilled practitioner adequately deals with in only a few seconds inevitably become

Herculean in size. Bloch's perspective of "connectionism," which postulates a different "cognitive mechanism" (multiple parallel processors) fails, however, to account for bodily knowledge and communities of practice, emphasizing instead knowledge as existing in the brain.

The perspective of intuitive knowledge and situated practice represents a radical challenge to cognitive anthropology and its preoccupation with intellectualized static systems. Few people actually operate in everyday life in the idealized manner described for the European navigator and the "rational" Alaskan skipper, and those who do are unlikely to do so on a regular basis. Many people act, much of the time, on the basis of intuitive knowledge and bodily dispositions – "tacit" knowledge (Polanyi 1958) or *techne* (Marglin 1990). As Suchman (1987: ix) suggests, "plans are best viewed as a weak resource for what is primarily *ad hoc* activity. It is only when we are pressed to account for the rationality of our actions, given the biases of European culture, that we invoke the guidance of a plan." To adequately represent situated practice, the flow of unfolding activity, and how we find our sea legs, other models are needed. And this brings me back to differential fishing success, Icelandic fishing, and the notion of the skipper effect.

9.3 Differential fishing success

In fisheries generally, catches vary from one boat to another. As Sahlins (1972: 73–74) observes, "for certain forms of production, notably hunting and fishing, the likelihood of differential success is known to common sense and experience." The nature and causes of such differences, however, are matters of some disagreement among anthropologists. In the literature on fishing, there has been a tendency to emphasize the variability of the skills of the leaders of fishing operations. Reviewing the literature, Acheson (1981: 290) points out that "the vast majority of anthropologists are convinced that . . . differential success is primarily due to marked differences in fishing skill." At the same time, several ethnographies have drawn attention to the social world to which the skipper belongs. Some have focused on the importance of the crew for skippers' decision-making and, by extension, fishing success – a classic example being Barth's (1966) analysis of the role of trust and self-confidence in Norwegian herring fishing (see also Heath 1976). Some accounts, too, have emphasized that the world of the ship should not be presented as an "endogenous" one, as a "closed cultural system" (Byron 1986: 96). In the latter view, any study of what happens on board fishing vessels must consider the role of fleet behavior and the larger contexts of fisheries and fishing communities. To what extent, then, does the focusing on the personality and production value of the fishing skipper contradict the emphasis on communities of practice (crews, fleets, and fisheries) and the situated nature of learning and decision-making?

When explaining differential fishing success, Icelanders often assumed, during most of the twentieth century, that the contribution of the leader of fishing operations overshadows that of everything else. Catches were said to vary from one boat to another during any one fishing trip, a single season, or a generation largely because skippers are different. In competitive fishing, the skipper's position or "seat" (*sæti*) in the local hierarchy of catches was a central concern; if the skipper had a low position, he risked losing his job. Once during my fieldwork, a skipper was fired mid-season by his company because "he did not fish enough." Being at the head of the fleet for the season, on the other hand, being *aflakóngur* (literally, "catch-king"), brought a high degree of honor and prestige. One of the central points in Icelandic theories of fishing success, then, was the attribution of particular qualities to the "good" skipper which enable him to catch more fish than others. The chief quality of a good skipper was held to be a very personal one, perhaps comparable to "intelligence" in psychological testing, a quality that is independent of the frequency of its application. For Icelanders such a distinction was just as important as the distinction made by many educators between the "cleverness" of a student and how often he or she puts it to work, the difference between intelligence quotient and diligence. The dominant Icelandic theory of fishing success, then, was *hierarchical* in that it emphasized differences among a group of producers, normally within a local fleet. As I shall argue later on, however, another (and somewhat less visible) element of indigenous discourse also drew attention to the co-operation of the crew and the wider context of communities and fisheries.

In Iceland, the hierarchical model of fishing success first developed at the beginning of the twentieth century, when the domestic economy gave way to entrepreneurial fishing and large-scale capitalist production for an expanding foreign market. In order to appropriate fish, skippers had to become fishers of men. Later – especially after the introduction of a quota system in the cod fishery in 1983, a few years after my original fieldwork – a new discourse developed in competition with the earlier one. The present system of quota management allocates a given share of the annual catch, a transferable quota, to each boat-owner. Fishermen and managers agree that, as a result, the "hunting element" of fishing (*veiðimennska*) has been disappearing. Sometimes they expressed the view that, with increasing governmental control of the industry, the custom of awarding the most successful skipper of the year a particular prize on Fishermen's Day (a standard custom in most Icelandic fishing communities for decades) was a little archaic; as fishing was "reduced" to business transactions, success became less a matter of fishing skills than capital and economics. The top skippers were simply privileged quota-kings (*kvótakóngar*).

Keeping in mind the importance of the hierarchical model in both the ethnographic literature and the Icelandic context, Durrenberger and I tried to separate the rhetorical element of production discourse and the realities of differential success, including the role of the skipper effect (Palsson and Durrenberger 1982, 1983). We tried to assess statistically the skipper's contribution in the case of the winter fleet of Sandgerði. The "residual" in our analyses, the percentage of variance unaccounted for by the variables of boat size and number of trips, represented the effects of all other factors that could possibly influence catch – such as the skipper and crew, the mechanical condition of the boats, the weather, and economic and political factors. Given the small residual in our calculations, we concluded that the skipper effect was far less than generally assumed. If differential success was principally a "material" matter, individual differences in fishing skills ("fishiness," *fiskni*) were overstated in folk accounts.

Several studies have addressed similar issues, attempting to identify the nature and causes of differential success by statistical means (see, for instance, Thorlindsson 1988). One of the problems, however, in most, if not all, of the statistical studies involved is that they exhibit certain methodological flaws and weaknesses. As Gatewood and Mace (1990: 346) point out, the findings of many of these studies typically remain ambiguous because the key "variables" (skipper's skills, in particular) are relegated to the status of residual effects – in other words, they are not measured directly. Another problem is that some of the variables – skipper, boat size, and fishing effort – tend to be confounded; a good skipper tends to have a large boat and to fish frequently and it is difficult, therefore, to separate the effects of person, effort, and technology.

The central issue, however, is not so much a matter of statistics and methodological refinement; rather, it has to do with the conceptual issues that remain. Take, for one thing, the "confounded variables" of person and technology. The problematic nature of technology and its relation to the person is well captured in the Icelandic distinction sometimes made between those who fish "by skill" (*af lagni*; literally, with dexterity) and those who fish "by force" (*af krafti*; literally, with power). In the former case, typically involving an experienced skipper, dexterity and alertness to the tasks at hand are of primary importance; in the latter case, usually that of a novice, it is technology that counts. The skillful skipper attends to his fishing technology as if it were an extension of his person; the novice, in contrast, focuses on the gear itself – fetishizing his gadgets, making more trips, and using more fuel, often destroying more gear in the process. Where are we, then, to draw the boundaries between fishermen, their technology, and the environment? Polanyi (1958: 59) observes that as we become skillful practitioners we assimilate technology as *a part* of our own body. In the process of learning, he suggests, "we shift outwards the points at which we make contact with the things that we observe as objects outside ourselves"; as tools become "part of ourselves, the operating persons," we "pour ourselves into them and assimilate them as parts of our own existence. We accept them existentially by dwelling in them" (Polanyi 1958: 59).

Significantly, in Iceland experienced skippers often speak of knowing the details and the patterns of the "landscape" of the sea bottom "as well as their fingers." This indicates that for the skilled skipper fishing technology – the boat, electronic equipment, and fishing gear – is not regarded as an "external" mediator between his person and the environment but rather as a bodily extension in quite a literal sense. Thanks to such technological extensions the experienced skipper is able to "see" the fish, an otherwise invisible prey, and the landscape of the seabed – much like, given the necessary training, a blind person is able to walk with the aid of a stick. Thus, as Bjarnason and Thorlindsson (1993: 388–389) point out, "skippers operating comparable boats will make different use of them in terms of effort and efficiency." Such bodily extensions have their ethnographic parallels; skilled fieldworkers pour themselves into their pen or notebook (or computer), not so much focusing on them as dwelling in them. In the situated practice of fieldwork, our equipment, too, has a tendency to "disappear" (see Suchman 1987: 53).

More importantly, given the present perspective, the whole framing of the issue of differential fishing skills rests on a reductionist notion of skill and the autonomous human agent, much like sociolinguistic notions of speaking and communicative competence (Palsson 1991: 17). Assuming that enskilment is mainly the result of practical engagement with the environment, the notion of the autonomous skipper is a misleading one; whatever different contributors to the discussion of the skipper effect may have said about the *size* of the statistical residual, and about the amount of variance in catches explained by the skipper's skill, there has been a tendency to think of the skipper as a separate "variable" more or less independent of social context, and to remove his actions from the sphere of social relations. Thorlindsson, who argues for the *strength* of the skipper effect by comparing the hierarchies of the Icelandic herring fishery and international chess, is no doubt right in suggesting that in both cases "individual performance is . . . based on skill," (1988: 206), but the analogy between a highly individualistic game and fishing – equivalent to Saussure's famous analogy between chess and speaking (see Harris 1988) – tends to draw upon a culturally specific view of the craftsman and producer. When applied to fishing, the analogy of chess suggests a dubious model of the skipper, a person operating outside society, removing his actions from his relations with crew, other skippers and their crew, and the wider community to which he belongs. If those who emphasize the strength of the skipper effect tend to lose sight of the social nature of the tasks involved, then statistical analyses (including my own) that purport to show the *weakness* of the skipper effect, relative to folk accounts, likewise fail adequately to render the collective project of fishing.

Again, indigenous discourse provides a lead. While Icelandic fishermen and boat-operators frequently underline the role of the skipper, they also emphasize the importance of "having a good crew" (*að hafa góðan mannskap*). One trawler skipper I talked to used the analogy of music, referring to the crew as an instrument and the skipper as a piano player:

> A good player is useless if the piano is deficient or out of tune. Likewise, a first class piano doesn't make good music if the player is no good.

Skippers' frequent statements about the significance of "having a good crew" have something to do with modesty; a good skipper is unlikely to elaborate on his own performance or capabilities, since boasting would not be tolerated. Nevertheless, there is every reason to take seriously skippers' claims about the importance of the crew; such statements are not merely rhetoric. To adequately render the social nature of production we may have to extend the notion of the "skipper effect" and refer as well to a "crew effect" (Rob van Ginkel 1992, personal communication) – if not to a "fleet effect" (White 1989). If, for the skipper, the crew is an "instrument" analogous to his boat and fishing gear, it logically follows, given what I said above, that it, too, forms a part of the operating person of the skipper. To return to the question I posed above, concerning the relative contributions of the individual and the larger social context, fishing success is both a matter of personalities *and* collectivities. And this is precisely to decenter the notion of skipperhood; in order to have success in catching fish, the skipper must dwell with his crew and the larger community of practice to which they all belong.

To continue the ethnographic analogy, to find their sea legs – to learn to "make good music," to use the skipper's words – ethnographers must pour themselves into the community they work in. While for anthropologists such a perspective of dwelling and empathy strikes a familiar note, it also runs against much accepted wisdom, including the current image of the lonely fieldworker and the alienating discourse on violence and boundaries associated with the notion of anthropology as cultural translation.

9.4 The flow and momentum of fishing

While fishing the crew performs a highly complex operation, involving a series of specialized and interrelated tasks, each of which must be performed quickly and in accordance with a given schedule. Fishing with gill nets is a good example. A net boat has several units of gill nets (*trossur*) and on each trip the skipper will normally visit several locations, drawing the nets set during the previous trip (usually a day or two earlier) and re-setting them. Deciding on the exact location is the skipper's responsibility; the crew are not directly involved although they take an active interest (given the application of a "share system") and, occasionally, especially when fishing is not going well, they may offer their own views. In contrast, setting the nets and drawing them are intensely co-operative tasks. Once the boat is in the right spot, the engine is set to full speed while the crew "cast" the nets overboard in a swift operation. The key issue here is to make sure that the nets do not become entangled, so that the operation can be completed smoothly and without accident.

Drawing the nets is a slower but far more complex process. The skipper oversees the whole operation from his base on the "bridge." Each member of the crew occupies a specific position on the deck, the most critical one being that of the operator of the winch that draws the net. Not only is this a highly dangerous task, it sets the speed of the operation and ensures the flow of the job. The position of the winch-operator tends to be permanent and specialized; other deck-hands, in contrast, rotate relatively freely from one position to another. Clearing the net and removing the fish also demands quick hands, careful attention, and good co-ordination. Each person has constantly to adjust his or her actions to those of the other crew as well as to the speed of the net passing through their hands. If one person fails to cope with the demands of the task, the whole operation comes to a halt. Keeping stops to a minimum not only diminishes the likelihood of accidents as well as maximizing free time ashore between trips, but also increases catches relative to effort (that is, per trip). Asked about the importance of co-operation among the crew, a manager of a large fishing firm (and a former successful skipper) explained:

> It almost makes *all* the difference to have a good crew. After all, a skipper cannot achieve considerable success unless he has a good crew. Two of our boats are almost identical, but in one case there is a constant turnover of crew – as three or four deck-hands leave the boat after each fishing trip – while in the other the crew remains essentially the same. The latter boat, with its stable and experienced crew, enjoys much greater fishing success.

Since an experienced crew is the key to efficiency and safety on deck, as well as to the economic use of gear and technology, boat-owners try to avoid a high turnover in personnel. Skippers, similarly, emphasize that successful fishing is only possible as long as certain interactional demands are met – a low turnover of crew and good relationships among the persons involved. Thus skippers at the top of the local hierarchy tend to have stable crews, while those at the bottom must accept inexperienced novices much of the time. Generally, a high turnover of crew is regarded as a bad sign with severe implications; it suggests low fishing effort (relatively few trips), limited catches in any one trip, low salaries, low morale, lack of prestige, and a high accident rate. Skippers point out that while every skillful deck-hand was once a novice, and while they may have to put up with one novice at any time, a "mixed crew" with a relatively high number of inexperienced deck-hands is both a safety hazard and a social and economic handicap.

Given the importance of skilled deck-hands and a low turnover of personnel, crew membership, the result of complex negotiations, needs careful consideration. Whether or not he owns his boat or works for someone else (the larger vessels are usually company-owned while the smaller ones tend to be the property of skippers and their families), the skipper is responsible for hiring personnel – one or two mates, a cook, one or two mechanics, and several deck-hands, depending on the size of the boat and the kind of fishing gear used. Potential crew members are likely, of course, to keep in mind the skipper's record in fishing. It is not, however, the only relevant consideration. Some skippers are known to be arrogant, strong leaders while others have the reputation for being democratic "nice chaps." As one deck-hand put it:

> Skippers very much differ from one another. A skipper may be good at catching fish (*aflamaður*), but he can be so boring that one is not willing to put up with him and work for him.

"Cultural" factors, therefore, may outweigh economic ones. Skippers, too, may be motivated by non-economic considerations. Some skippers believe that women pose particular difficulties at sea irrespective of their competence in the narrow context of fishing. Also, friendship and other personal connections may be more important to the skipper than the reported "market" value of a deck-hand. Often, however, it is difficult to separate the economic and the cultural; in the long run, cultural considerations may *become* economic, as self-fulfilling prophecies.

A good skipper needs a disciplined crew, but he also needs to be a friend of those who work with him. During fishing trips, the relationship between skipper and crew tends to be authoritarian and rigid, but when they reach harbor they often "loosen up," shifting to a more informal gear. Sometimes skippers socialize with their crew between fishing trips, strengthening social bonds and enhancing the solidarity of the crew, even though such occasions lack many of the ritual aspects Johnson (1979: 248) describes for the "group-binding" Portuguese *caldeirada* (a joint meal at a local tavern). While the hierarchy of skippers is subject to some stability, the skipper's trust and reputation is not given once and for all from the moment he begins to fish, as Barth's account would lead one to believe, emphasizing the "self-confirming pattern" (1966: 10) whereby "good" skippers get stable crews, fish independently, and land high catches (cf. Heath 1976); rather, trust and reputation are negotiated and maintained in the context of social discourse and changing social and economic realities, in the flux and movement of boats, capital, skippers, and crews.

The fishing crew, of course, is part of a larger context. Fishermen often speak of the personnel (*mannskapur*) of a boat in an extended sense – including several people ashore, those who ensure efficient repairs of equipment between fishing trips and those who bait lines and take care of nets, repairing old ones or supplying new ones. Indeed, folk accounts of fishing success often emphasize the importance of good fishing gear and the diligence of the people ashore responsible for its maintenance. "Having a good crew," therefore, means not only being able to rely on a good *fishing* crew, but also being provided with good "services" (*þjónusta*) on land. Other ties connect the fishing crew to a series of social networks ashore, family and relatives, neighbors, and friends. Such networks are not limited to the landing port in question as they extend over a wider area. Skippers and crew get together in their respective regional organizations and unions, to negotiate shares and salaries and, generally, to defend their economic interests. Such relations significantly affect the process of recruitment, the mutual knowledge of fishermen, working conditions on board the boat, and, generally, the momentum of fishing.

The fleet is ever-present as well. While one may speak of the vessels temporarily associated with a particular landing port and nearby fishing grounds as a "local" fleet, such a fleet knows no clear boundaries (cf. Acheson 1988). The fleet, in fact, is a changing constellation of boats that are registered in different towns and municipalities, and many of the skippers, crew, and boat-owners involved are permanent residents of other localities. Moreover, the fleets of different ports are hard to separate; during fishing they merge on the boundless sea. Nevertheless, the communion at sea is a very important one. Inevitably the skipper's decisions while fishing are constrained by the decisions of other skippers and by the movements of the fleet. While deciding where to fish is largely guided by the readings of electronic equipment and by the skipper's experience of earlier fishing seasons, of no less importance is knowing what *other* skippers are doing, where they are likely to be, and how much they will catch. There are obvious benefits in co-operating with other skippers and crews on a daily basis while at sea, especially after a long break in fishing; sharing information on the state of the major areas saves time and fuel and each skipper gains information about fish migrations that he could not acquire on his own. As Wilson (1990: 14) points out, while fishermen seek to reduce the search problem by looking for recurrent patterns in the migration and location of fish, "the number of observations necessary to establish regularity is far too large for any single individual to acquire." White (1989: 25) proposes the notion of the "fleet effect" to address this issue, suggesting that information-sharing and variable discovery techniques within a fleet "result in larger catches than if all boats fished alone." However, because skippers compete among themselves for locations, fish, crew, and prestige, they often carefully guard valuable information available to them.

One way to solve the tricky problem of co-operating under conditions of competition is to participate in an informal club of skippers, a network with relatively stable membership, thereby exchanging information on a regular and reciprocal basis: "This method of trading information tends to maintain strong incentives for the acquisition of new knowledge and at the same time tends to coordinate and disperse information in a way that provides few benefits for free riders" (Wilson 1990: 15). Icelandic skippers refer to the participation in clubs of this kind as "belonging to a code" (*vera í kóda*); club members agree on a secret code or language which enables them to freely discuss important issues (especially catches and locations) on the inter-boat radio (a device fishermen call "the spy") without providing information to "outsiders." In the absence of direct, visual clues, skippers try to learn as much as possible about their colleagues (whether club members or not) through the use of the inter-boat radio. The lengthy and repetitive exchanges between skippers, and the gossip, teasing, and joking they contain, often seem to be simply an end in themselves, but one should not underestimate their social significance. Such exchanges not only maintain communion in the fleet, they also provide subtle information on fishing. What skippers say is interpreted on the basis of an extensive prior knowledge of catch records, fishing locations, and fellow fishermen. With the introduction of mobile telephones and the Internet, which allow for increasing privacy in communication between club members, the role of the inter-boat radio has been much reduced.

Due to the physical seclusion of the crew while at sea, the competitive nature of fishing, and the changing and somewhat arbitrary definition of "local" boats, it is easy to lose sight of the importance of the fleet. During my early fieldwork I was suddenly reminded of the sociality of the fleet by a search of an unusual nature – a tragic event in which a small boat with two men on board disappeared in the middle of a fishing trip, apparently due to rough seas. Quite abruptly, "the fleet" became visible, operating as a single unit, scanning the seascape for several hours – searching, without success, for the lost boat and crew.

The reference to sociality is not to suggest that crews and fleets are best described as collective, rule-governed entities – as sets of "markedly stereotyped, detailed roles" (Barth 1966: 10) generated on the basis of formal statuses and transactional constraints. As I remarked earlier, with reference to Rosaldo's example of the game of basketball, rules are not the essence of games from the point of view of those who play them. Nor am I arguing that crews and fleets are supra-individual byproducts of interaction and co-ordination, reified entities in the Durkheimian sense. If we were to follow the Durkheimian lead, the agency of the skipper (and any other member of crew, of course) would dissolve in a hierarchically more inclusive agency of the collectivity of crew and fleet. Given an interpretation that abandons any radical distinction between the individual and the collective, assuming that each person is an "ensemble" of social relations, as Marx once put it, the skipper is endowed with both agency and sociality. As developmental psychologists Fischer, Rotenberg, Bullock, and Raya (1993: 97) point out in their discussion of learning, "the skill concept includes the person as well as the context. It is as much a mistake to leave out the person as to leave out the context." A skipper's agency is inevitably constituted within the nexus of social relations, and yet his personal authority is essential for the collective project of fishing.

The collaborative mode of practice and enskilment emphasized in the present account is increasingly the subject of anthropological discussions, usually in connection with the notion of "apprenticeship" (see, for example, Coy 1989). Many accounts of *ethnographic* practice, however, are excessively self-indulgent and individualistic, in the writing-culture genre. Assuming the perspective of practice theory, and rejecting any radical distinction between experts and laypersons, it seems essential to extend the notion of apprenticeship to the work of anthropologists – and this means to move from autobiographies and narcissistic accounts of autonomous selves to social histories of fieldwork as a co-operative enterprise sustained by relationships with other people (spouses, friends, informants, and colleagues). Fieldwork, as Gudeman and Rivera emphasize (1995), is teamwork with a flow and momentum of its own – the anthropologist produces his or her ethnography *with* a responding people; the fieldworker is an apprentice in the collective enterprise of making an ethnography.

9.5 Apprenticeship and attentiveness

Becoming a skillful skipper – a competent member of the collectivity of a fleet – demands several years of training, both formal and informal. A prospective skipper, usually a fisherman's son, begins his career as a deck-hand, often in school holidays during early adolescence. To get a skipper's license, however, a fisherman has to receive formal training in a specialized institution, the Marine Academy. The formal title of the "skipper" (*skipstjóri*) assumes particular legally enforced rights and obligations, emphasized and assimilated in the classroom, through two years of formal instruction. Much of the training focuses on navigation, safety measures, and the use of electronic equipment. For skippers, however, enskilment in *fishing* is not a matter of formal schooling and the internalization of a stock of knowledge; rather, it is achieved through active engagement with the environment, in the broadest sense of the term. As one skipper explained,

> the Marine Academy teaches interesting subjects, including meteorology. On the other hand, you learn even more by simply taking part, by living the life at sea – especially if you are young and enthusiastic.

"Real" schooling, then, is supposed to take place in actual fishing. The emphasis on "outdoor" learning is emphasized in frequent derogatory remarks about the "academic" learning of people who have never "had a pee in a salty sea" (*migið í saltan sjó*). Questioned about the role of formal schooling, skippers often say that what takes place in the classroom is more or less futile as far as fishing skills and differential success are concerned, although they readily admit that schooling has some good points, preventing accidents and promoting proper responses in critical circumstances involving the safety of boat and crew. Even a novice fisherman, they say, with minimal experience of fishing, is likely to know more about the practicalities of fishing than the teachers of the Marine Academy. The most important items of equipment, including fish-finders, continue to be modified, new models constantly enter the market, and, in any case, each tool or machine has its "nuances" that need to be studied and established in context, in the course of fishing. No formal training can cope with this flexibility and variability in the real world. Therefore, there is little, if any, connection between school performance and fishing success.

Advice by former skippers may be useful schooling, but here, again, what counts is what the skipper does with his information on the spot, during actual fishing. On one occasion I witnessed a discussion between a novice skipper and a retired one about a "rocky area," a particularly difficult place to fish. The experienced skipper advised the young one on how to maneuver the boat and the fishing gear, elaborating on the complex details of currents, fish migrations, and seabed features. The novice listened carefully and then commented, somewhat perplexed, "I guess I'll figure it out when I get there." Some years earlier the young skipper had been a deck-hand on the old man's boat. That job had different demands – relating more to what happened aboard the vessel than in the sea – and, therefore, he said, he had rarely "attended to" (*fylgdist með*) the task of locating and catching fish.

Skipper education recognizes the importance of situated learning. Earlier participation in fishing, as a deck-hand (*háseti*), is a condition for formal training, built into the teaching program; this is to ensure minimum knowledge about the practice of fishing. Once the student in the Marine Academy has finished his formal studies and received his certificate, he must work temporarily as an apprentice – a mate (*stýrimaður*) – guided by a practicing skipper, if he is to receive the full license of skipperhood. The attitude to the mate varies from one skipper to another; as one skipper remarked, "some skippers regard themselves as teachers trying to advice those who work with them, but others don't." While skippers differ from one another and there is no formal economic recognition of their role in this respect, in terms of a teaching-salary, according to many skippers the period of apprenticeship is a critical one. Reflecting on his mentor, with whom he had spent several years at sea, one skipper explained: "I acquired my knowledge by working with this skipper, learning his way of fishing. I grew up with this man." It is precisely here, in the role of an apprentice at sea, that the mate learns to attend to the environment *as a skipper*. Working as a mate under the guidance of an experienced skipper gives the novice the opportunity to develop attentiveness and self-confidence, and to establish skills at fishing and directing boat and crew. The role of the mate, in fact, institutionalizes what Lave and Wenger (1991) term "legitimate peripheral participation," a form of apprenticeship that allows for protection, experimentation, and varying degrees of skill and responsibility. This is not a one-way transfer of knowledge as the skipper frequently learns from the co-operation of his mate; mate and skipper – in fact, the whole crew – educate each other. In the beginning, the mate is just like an ordinary deck-hand; in the end he is knowledgeable enough to have a boat of his own. At first he is of little help to his tutor, but later on he can be trusted with just about anything; occasionally, the skipper may even take a break and stay ashore, leaving the boat and the crew to his mate.

Often the advice of the skipper is in the form of verbal directions. He will draw the novice's attention to various aspects of skipperhood – how to maneuver the boat, how to use electronic equipment, how to follow fish migrations, and so on. A skipper is unlikely to share his most personal tricks, as the mate may later on become one of his competitors, in charge of another local boat. Every skipper has a personal "diary" with details about times, species, catches, and fishing locations, and mates are rarely, if ever, given *direct* access to this information (some of it is stored in the memory of a computer located in the wheel-house and, significantly, passwords are sometimes used for protection). On the other hand, a keen novice will gradually learn to imitate the actions of his skipper, observing his decisions and carrying out his commands. One skipper pointed out that he "simply" learned most of what he knew "by *seeing* how others handled their tasks."

From the point of view of the mate, the choice of a tutor is very important. In the words of one skipper, "to get an experienced and clever man is the most important thing that can happen to a beginner, the greatest luck of all." Some skippers envy colleagues who have had the opportunity to work early in their career with renowned skippers, acknowledged mentors and "men of catch" (*aflamenn*). Sometimes, they say, the period of apprenticeship is terminated too soon. Thus, one skipper who enjoyed modest success claimed that he "did not spend enough time as a mate" with an experienced skipper, and that he regretted having left too soon to become a full-time skipper. This was the main reason, he continued, why he could not compete with the top skippers. There may be several reasons, personal and economic, for quitting apprenticeship too early. For one thing, a retiring skipper-owner, or someone with a temporary health problem, may encourage his son to take over at an early stage, before the latter really matures as a skipper. This seems to have been the case with a young and unsuccessful skipper in one of the communities in which I worked. He was repeatedly subject to critical discussions and some even laughed at his flamboyance and inexperience; one fisherman commented, when hearing the skipper's voice on the inter-boat radio, that the skipper "could not even handle the channels on the radio." Another fisherman made sarcastic comments about the "unlucky" skipper who tended to go fishing in bad weather and vice versa, staying home when the weather was good. The implication was that the skipper was inattentive. Other skippers, it seems, were reluctant to respond to the novice on the radio, despite his insistence to communicate. Skippers who do not take the period of apprenticeship seriously, or who cut it short, invariably fish "by force" and not "by skill"; they are referred to as *göslarar*, inattentive and clumsy skippers. To become a respected skipper, a fisherman not only has to become a mate and later on a skipper, he also has to "prove himself" (*sanna sig*) in his capacity as skipper over several seasons, "growing up" or "developing himself" (*verða að manni*).

The skipper's knowledge is a complex one; a skipper must choose times and places to fish on the basis of a series of detailed environmental information. It is not surprising, therefore, that fishermen often refer to the importance of "attentiveness" (*eftirtekt*, *athygli*) and "perceptiveness" (*glöggskyggni*); the ability to recognize and apply an array of minute but relevant details. Attentiveness is a complex ability and includes, for example, being able to "read" the sky and predict the weather, to participate in discussions within the local fleet, to understand the "sparks" of electronic instruments, and to be able to co-ordinate crew activities. In some fisheries, technological and economic changes seem to have resulted in rapid de-skilling. There is little reason to believe that this has been the case in Iceland. While old and retired skippers sometimes point out that fishing has been radically transformed by electronic technology (including the computer), emphasizing that "natural signs" are increasingly redundant, attentiveness continues to be one of the central assets of the good skipper and, just as before, it demands lengthy training. The skipper's universe is very different from that of his colleagues of earlier decades, but what shows on the screens of the radar, the computer, and the fish-finder is just as much a "natural sign," directly sensed, as birds in the air or natural landmarks.

As in the Alaskan situation described by Gatewood (1983), the most important test of skippers' skills, and the most relevant criterion for their prestige, is fishing success. Good grades and formal schooling – and sizable fishing quotas – are useless if one fails to catch fish. Unlike their Alaskan colleagues, on the other hand, Icelandic skippers are not expected to follow "rational" procedures, nor do they say they necessarily do. If skippers account for their procedures at all when making decisions about the locations of prey, they are likely to say that they follow what Gatewood calls "reasonable" procedures. Many skippers have described how a dilemma regarding fishing locations was solved by a strange message or intuition, some kind of "whisper." When explaining exceptional catches during single fishing trips, they sometimes claim they simply know intuitively (*finna á sér*) what to do. Often they also refer to "messages" or hunches received on the spot and the state of "fishing mood" (*fiskistuð*), a peculiar psychological experience – "like being possessed" – beyond skippers' understanding and control. As one skipper pointed out,

> sometimes one is so thoroughly confused when leaving harbour that one hasn't got a clue as to where to go. Then one may suddenly get a hunch.

Skippers sometimes speak of cerebral processes in this context, of "knowledge in the head" and "being able to think like fish," but they also refer to wisdom "carried in the blood" and "unexplainable" bodily judgment.

Skippers are likely to have a plan when they embark on a fishing trip, but in the process of fishing they may be forced to revise it, or even to abandon it:

> Sometimes one tries to follow some plan, based on one's experience from the day before or the recent past. And then a new day arrives, with new kinds of circumstances, and it simply becomes impossible to follow a plan.

While fishing, the skipper's tactics are continuously adapted to new circumstances – the teamwork on board, the decisions of other skippers, the amount and kinds of fish caught, weather forecasts, and ocean movements.

Although skippers frequently discuss the relevance of attentiveness, they rarely mention how they actually make decisions. One reason is that they are guided more by practical results than by an interest in theoretical advancement. Often they "simply" notice that a particular strategy seems to work, without worrying about why that is the case. Skippers' reluctance to discuss their own fishing tactics is not only the result of competition and secrecy, and the value they generally place on independence and modesty; they have difficulties in verbalizing their complex experience and intuition. What fishermen label as hunches and fishing mood is particularly difficult to verbalize; some important decisions are made "out of the blue." What matters most of all, one skipper explained, is "how one's body will cope." Decision-making, then, is based less on detached calculation or "mental" reflection than on practical involvement. Similarly, the process of enskilment is not just a cognitive one; rather it involves the whole person interacting with the social and natural environment.

Once again, the analogy with ethnographic practice is revealing. Gatewood (1985) has told the story of how he became a competent deck-hand in salmon seining in Alaska. After only a few days he had to abandon his "arrogant" hopes of being a participant-observer; trying to cope with the demands of fishing did not leave much time and energy for anything else – "I had all I could handle," he says, "trying to participate" – and besides, the cognitive demands of "observation" were quite irrelevant, if not detrimental, to the learning process (Gatewood 1985: 205). Learning was not a matter of verbal encoding and enculturation but of adapting to the "practical constraints" of the job:

> The development of a seiner's cognitive organization is directed not by some transcendental need to share meanings but by the practical constraints of coordinating his actions with those of his fellows. His actions, and theirs, speak louder than words. (Gatewood 1985: 216)

Edelman's (1993) ethnographic study of shunting in a Swedish railway yard provides similar observations. At the beginning of her fieldwork Edelman was preoccupied with the "middle-class" notion of personal autonomy, of distinguishing herself from others, a notion which proved to be an obstacle to the progress of her fieldwork. Later on, however, she was "beset . . . by the reciprocity of the demands and obligations that spring from the daily toil, the constant efforts to co-operate and create a 'flow' in the work, the logic of the work-process, and the meanings that were attached to individual actions" (Edelman 1993: 160–161).

As a novice fieldworker and fisherman I, too, could not help noticing the importance of co-operation, of co-ordinating one's bodily movements and the actions of others to ensure the continuity of the enterprise. Such mutual attentiveness, the result of collective enskilment, is essential for efficient teamwork, the synchronization of the tasks involved. Occasionally, the Icelandic skipper will shout commands at his crew, over the noise of the engines, during the process of drawing a net: "Loosen the net!" or "Stop the winch!" The better the crew know each other, however, the more likely it is that they will be able to predict one another's actions and respond appropriately in emergencies, and the less likely is the skipper to intervene. Silence is a sign of smooth co-operation and, likewise, direct commands signify the opposite – low morale. Most of the time, words are beside the point simply because actions speak *faster* than words.

9.6 Conclusions

Icelanders, as we have seen, sometimes apply the metaphor of the journey, the fishing trip, to the issue of personal enskilment. This is to suggest that learning is not a purely cognitive or cerebral process, a mental reflection *on* differences in time and space, but is rather grounded in the contexts of practice, involvement, and personal engagement. Enskilment is not only *likened* to the physical experience of seasickness, it is indeed a bodily exercise. To become skilled at something like catching fish is to progress from nausea to well-being, to feel at home in both one's body and the company of others. As Bourdieu reminds us, socialization produces very real, physical effects, durable dispositions inscribed in the *habitus*. The novice imitates the *actions* of others, not simply their models and discourses; practical schemes "pass directly from practice to practice without moving through discourse and consciousness" (Bourdieu 1990: 74).

Another argument from Icelandic discourse, concerning the situatedness of decisions and the importance of teamwork, poses a fundamental challenge to the terms of academic debate on differential success, in particular the "skipper effect." Fishing, after all, is a social enterprise. While the skipper is unlikely to acknowledge, let alone ask for, the advice of his crew and other skippers, and although it may be tempting to think of his decisions as independent ones and more or less his own, each of them is constrained by the context in which it is made, much like a discourse is inevitably constrained by context, including what has been said before. In fact, we may well think of a fishing season as a "long conversation" involving a whole community of actors – both at sea and ashore – in which the agency of each is continually constituted in relation to the others. Folk accounts of the context of decision-making and the importance of the crew and the local fleet indicate an indigenous practice theory of fishing, emphasizing that what skippers do and what happens at sea is not the result of an internalized system of formal decision rules, as normative learning theory assumes, but of hunches and tacit knowledge, of insights developed on the spot but firmly grounded in the flux and collectivity of fishing.

For those familiar with the mood and momentum of a game like basketball – the rhythm of a team and the flux of physical movement and bodily contact – the practice perspective of fishing immediately has parallels. As Larry Bird, a superb basketball player, reasoned, "[a lot of the] things I do on the court are just reactions to situations . . . I don't think about some of the things I'm trying to do . . . A lot of times, I've passed the basketball and not realized I've passed it until a moment or so later" (cited in Dreyfus 1991: 93). Skills – in fishing or doing fieldwork (or anything else, for that matter) – are indeed individual in the sense that they are properties of the body, dispositions of the *habitus*. However, to isolate their acquisition and application from everything outside the boundaries of their soma is to subscribe to a normative theory of learning and a natural conception of the individual. An alternative approach recognizes the sociality of the individual being and the situated nature of human activities. If, as Bakhtin (1986: 293–294) has argued, every word in conversation is half someone else's, every fish that gets caught is partly that of others.

It is one thing to participate in face-to-face interaction and to engage in fieldwork and quite another, perhaps, to "write it *up*." What would the perspective of practice theory adopted here imply for the latter project – the construction of comparative knowledge or ethnographic writing *outside* the field? Bloch (1991: 193) emphasizes that anthropologists must be aware that by writing up the knowledge of the people they study they are not merely reproducing it, they are also "transmuting it into an entirely different logical form." How can we, then, textualize the knowledge we acquire in the field without – in the process – fundamentally distorting both its character and the manner in which it was acquired? The important issue is not, however, whether or not our models agree with the models of the actors, but the way in which we *continue* the ethnographic conversation, the character of the social relations of ethnographic production. It may be easy to labor under the Orientalist illusion – assuming, as Knowlton (1992: 78) puts it, that we are able "to keep the 'native' safely in our notes or at the end of a journey, rather than among us" – since, thereby, "we avoid having our caricatures challenged and our categories confused in the messiness of solidary sociality." By terminating the ethnographic conversation once we are "out of the field," we preserve our essentialist enterprise, localizing our anthropological voices rather than anthropologizing local voices. Arguably, however, we never actually leave the field as long as we take part in the ethnographic enterprise; in one way or another we keep on conversing with the people we study, either directly (especially nowadays) as anthropologists "at home" or as visitors returning to a host, or, indirectly, through inner talk.

If the act of writing up is inevitably situated in a boundless community of practice – a community that admits no radical distinction between modelers and participants, scholars and natives, scientists and informants – then indeed ethnography is best regarded as a "perpetual discussion" (Gudeman and Rivera 1990: 4). Just as it is important to go beyond the "claustrophobic" concepts of cognition characteristic of normative learning theory (Lave 1988, Reed 1993: 63), it is important to go beyond the monologic notion of ethnographic production. In such a view, ethnography becomes, to paraphrase Benhabib (1992: 9), a "moral conversation exercising enlarged thinking." As ethnographers, we may be nauseated at times, given the shifting grounds of postmodernity and the contemporary critiques of monopolistic narratives and totalizing accounts. However, only by engaging ourselves in a continuous dialogue – with both our "informants" and fellow ethnographers – can we realistically expect to find our ethnographic sea legs.

10 Environmental relations: political economies

Much anthropological thinking, in different academic settings, representing a wide range of theoretical "schools" or paradigms, assumes a fundamental distinction between nature and society. Hollingshead, whose ideas influenced the cultural ecology of Julian Steward, expressed one formulation of such a dualism in clear and simple terms, speaking of "the ecological and sociological orders":

> The former is primarily an extension of the order found everywhere in nature, whereas the latter is exclusively, or at least almost, a distinctly human phenomenon ... The ecological order is primarily rooted in competition, whereas social organization has evolved out of communication. (Hollingshead 1940: 358)

Dualist theory, it was generally assumed, was on the right tracks: "[N]ow that the problem is recognized and a beginning made," Hollingshead (1940: 358) suggested, "we may expect a solution." Such a theoretical beginning was reinforced by a rigid academic divison of labor and massive institutional structures. The sociological order remained the subject of anthropologists and sociologists while the ecological one belonged more properly to professional ecologists.

Having established a fundamental dichotomy, Hollingshead, and many of those who followed him, usually qualified the dualistic thesis, emphasizing that nature and society were not to be seen as *totally* separate spheres but dialectically interlinked; each order "complements and supplements the other in many ways" (Hollingshead 1940: 359). Modern-day ecologists continue to "compare" the orders of nature and society as if they were separate, autonomous systems, exploring the links between them (Holling, Gunderson, and Peterson 1994). Despite the dialectic, interactive language, then, the boundary between society and nature remains a contested interface. During much of the twentieth century social theorists intensely debated the relative merits of two kinds of determinisms, the "prison houses" of language and naturalism. In the 1970s, Sahlins quite suitably characterized anthropology, a discipline continually trapped between idealism and materialism, as a "prisoner pacing between the farthest walls of his cell" (1976: 55), reinventing the allegory of the cave from Plato's *Republic*. In recent years, however, the weary debate of materialist and cultural reason has rather unexpectedly been replaced by a more fundamental one: The distinction between nature and society, one of the key constructs of modernist discourse, has itself increasingly been subject to critical discussion in several fields, including anthropology and environmental history. This development, partly a response to the postmodern, linguistic turn, global environmental problems, modern information technology, the greening of public discourse, and the redrawing of disciplinary boundaries, poses new challenges for social theory and ethnographic practice, setting the stage for a novel kind of environmental anthropology.

One possible avenue in that direction is to extend the Marxian approach, an approach usually restricted to *human* relations, to the analysis of human–environmental relations. Tapper (1988: 52) has argued that in hunting and gathering societies humans and animals engage in the "mutual production of each others' existence" and Brightman (1993: 188) similarly alludes to an "Algonquian labor process" in the case of the Canadian Cree, a process "in which humans and animals successively participate as producers of the other, the animals willingly surrendering the 'product' of their own bodies and the hunters returning it to them as cooked food, all figured in the idiom of 'love.'"

Drawing upon such perspectives, my aim is partly to show that similar discourses are applied to rather different theoretical contexts. Discourses on nature, ethnography, and translation, I suggest, extending arguments developed by Donham (1990), Bird-David (1993), and some others, often have much in common, notably the metaphors of personal relatedness and classic rhetorics. More generally, this chapter argues for the integration of human ecology and social theory, drawing upon perspectives often associated with Marx and Dewey, seeing humans in nature, engaged in situated, practical acts. I distinguish between three kinds of paradigms – orientalism, paternalism, and communalism – each of which represents a particular stance with respect to human–environmental relations. The paradigm of communalism differs from both orientalism and paternalism in that it rejects the radical separation of nature and society, object and subject, emphasizing the notion of dialogue. While ethical approaches to the environment and human–environmental relations are highly interconnected, I am less concerned with the former than the latter. Merchant (1990) has applied a taxonomy, similar to the one I am suggesting for human–environmental relations, to environmental ethics, distinguishing between egocentric, homocentric, and ecocentric approaches.

10.1 The political economy of the environment

The modern nature–society dichotomy is often taken for granted. It is necessary, therefore, to situate it in a wider historical and ethnographic perspective.

In medieval Europe, there was no radical separation of nature and society; if the dichotomy existed it must have been very different from that typical for the modernist project. As Gurevich (1992: 297) argues, in medieval times "man thought of himself as an integral part of the world ... His interrelation with nature was so intensive and thorough that he could not look at it from without; he was inside it." Significantly, the medieval term "individual" originally meant "indivisible" – that which cannot be divided, like the unity of the Trinity. The change in the meaning of the concept, the adoption of the modern connotation emphasizing distinctions and discontinuities, "is a record in language of an extraordinary social and political history" (Williams 1976: 133). The systematic fragmenting of the medieval world and the "othering" of nature it entailed first took shape in the Renaissance period, during which the whole Western attitude to the environment, knowledge, and learning was transformed.

The three-dimensional space established by Italian painters, including Sandro Botticelli, during the fourteenth and fifteenth centuries is one of the key elements of the epistemological revolution of the Renaissance. For early Renaissance painters, trained in the holistic world of Aristotelian philosophy and the medieval church, the canvas was primarily decorative space for the glorification of godly designs. By the end of the Renaissance, in contrast, the art of painting consistently focused on cognitive and spatial research, the representation of human activities and their place in nature and history. Renaissance painters were rewarded for their efforts with spectacular artistic success, the laws of perspective (*perspectiva*, or "seeing through"). In a brief period, nature became a quantifiable, three-dimensional universe appropriated by humans. This "anthropocracy," to use Panofsky's (1991) term, represented a radical departure from the enclosed universe of the Aristotelians constituted by the Earth and its seven surrounding spheres. The Cartesian anxiety of estrangement and uncertainty, however, of the separation from the mother-world of the Middle Ages and the nursing Earth, was compensated for by the rational ego, the obsession with objectivity, and a "masculine" theory of natural knowledge: "'She' [nature] becomes 'it' – and 'it' can be understood and controlled. Not through 'sympathy' ... but by virtue of the very *objectivity* of the 'it.' The 'otherness' of nature is now what allows it to be known" (Bordo 1987: 108).

If nature is an "Other," it has to be "translated"; much like the noise in the ruins of the Tower of Babel it demands close attention and effort at understanding. Such efforts, however, can take different forms. Students of literary translation emphasize that although translation may be seen as a perfect marriage between two different contexts, an important element in translation proper concerns the relations of power between "source" and "receptor" (Lefevere and Bassnett 1990). A translation indicates the relative submissiveness or superiority of the translator and the authority of the receptor vis-à-vis the source. Such a perspective may be applied to the ethnographic enterprise. How ethnographers, as visitors or guests, meet their hosts (and how they are met by them), how they manage their lives among them, and how they report what they experience, varies from case to case. Thus, one may speak of different relations of ethnographic production.

Similarly, emphasizing the contrast between domination and protection with respect to the environment, we may distinguish between two radically different kinds of human–environmental relations, environmental orientalism and paternalism. The key difference between them is that while the former "exploits," the latter "protects." Environmental orientalism suggests negative reciprocity in human–environmental relations, whereas paternalism implies balanced reciprocity, presupposing human responsibility. In the case of both environmental orientalism and paternalism, humans are masters of nature. Rejecting the radical separation of nature and society, object and subject, and the modernist assumptions of othering, certainty, and monologue, adding the dimension of continuity and discontinuity yields a third paradigm which may be referred to as communalism. This paradigm suggests generalized reciprocity in human–environmental relations, invoking the notions of contingency, participation, and dialogue.

Analogies of the human world and the natural environment need not be surprising. Humans often treat other human beings and the environment in a similar manner. Indeed, discourses on nature, ethnography, and textual translation have much in common. Thus the metaphoric language of classic rhetoric – of irony, tragedy, comedy, and romance – has appeared in a wide range of fields and contexts at different points in time. Donham (1990: 192) argues that even though the attempt to construct typologies with the "dramatic" metaphors of rhetoric "is bound to result in a certain crudeness, questions of rhetoric nevertheless appear to delineate ... the manner in which all social theories proceed from particular moral assumptions." Another metaphoric association draws upon the language of personal relatedness, of kinship and sexual relationships; such metaphors, as we will see, have often been used to represent both textual translation and the nature/society interface.

10.2 Orientalist exploitation

The paradigm of environmental orientalism not only establishes a fundamental break between nature and society, it also suggests that people are masters of nature, in charge of the world. If humans are not quite godly beings, at least they compete with God; thus the arrogant statement reported for Carl von Linné, the arch classifier of natural species, that "while God created Nature, *he* put it in order." The vocabulary of orientalism is typically one of domestication, frontiers, and expansion – of exploring, conquering, and exploiting the environment – for the diverse purposes of production, consumption, sport, and display. To the extent that one can speak of environmental "management" in this context, management is simply a technical enterprise, the rational application of Baconian science and mathematical equations to the natural world. This typically suggests a lofty stance with respect to the "object" in question. In the orientalist context, scientists present themselves as analysts of the material world, unaffected by any ethical considerations. This implies a radical distinction between laypersons and experts, another theoretical construct rooted in the innovations of the Renaissance.

Given the persistent othering of the object of modernist scholarship, the Baconian imagery of sexual assault, of "entering and penetrating ... holes and corners" (Francis Bacon, cited in Bordo 1987: 108), is a recurrent one. As Bordo (1987: 171) and Nelson (1992: 108), among others, have shown, the literature on modern science is replete with passages that describe human–environmental interactions by means of an aggressive, sexual idiom; nature appears as a seductive but troublesome female. Anthropology is not exempt from modernist, sexual jargon and predator–prey metaphors. Malinowski (1972: 8) argued, for instance, that

> the Ethnographer has not only to spread his nets in the right place, and wait for what will fall into them. He must be an active huntsman, and drive his quarry into them and follow it up to its most inaccessible lairs.

This is the rhetoric of the classic ethnography produced during the heyday of Western colonialism. Orientalist ethnographers colonize the reality they are studying in terms of a universalist discourse, asserting the superiority of their own society in relation to that of the natives. Given that anthropology was the offspring of colonialism, the predominance of the objectivist and orientalist extends over a long period in the history of the discipline. Textual translation has often been rendered in similar terms. Some leading students of translation talk about the relationships between translator and author not only in terms of a predator–prey relationship, they also tend to employ a violent sexual language. The content of the source-text is represented as a passive, female prey to be appropriated by a male translator.

Many examples of the industrial exploitation of "wild," undomesticated species illustrate the characteristics of environmental orientalism. The literature on fishing economies, for example, often attests to an aggressive stance; the expansive Icelandic fishing economy is one case in point. In the competitive fishing of most of the twentieth century, the chief criterion used for evaluating the social honor of a skipper was the relative size or the volume of catch, not the relative value of what was landed. The hero of fishing was the brave skipper who might risk the crew for extra tonnage. During this period, the sea represented a gigantic, continuous mass of energy to be worked upon actively and offensively by humans – more specifically, by daring males almost at war with the ecosystem (see Palsson 1991).

To capture the morality of evironmental orientalism and its impractical consequences, the rhetorical metaphor of irony may be a useful one. The producers naïvely expect to be in total control and yet by their own practices they seriously undermine their mastery, sometimes bringing the species they exploit to near depletion. To act in terms of concepts that have such unintended consequences is, indeed, rather ironic. Even more ironically, faced with the realities of resource depletion people sometimes adopt the fatalistic attitude that depletion is simply an inevitable ingredient of economic progress. The metaphor of irony, however, has probably enjoyed far less popularity, at least in academic circles, than the one of tragedy; witness the exponential growth in the literature on the "tragic" theory of the commons. Governmental authority or privatization, it is often assumed, are the only alternatives to individual greed and environmental abuse. In one sense, however, the orientalist regime has no drama at all; there is no environmental problem to solve, no need for corrective measures and scientific, ecological, or social expertise.

10.3 Paternalist protection

While the paternalistic paradigm shares some of the modernist assumptions of orientalism (it, too, implies human mastery and a distinction between laypersons and experts), it is characterized by relations of protection, not exploitation. This involves privileging scientific expertise, an inversion in the relative power of experts and laypersons. In the modern, environmentalist view, humans have a particular responsibility to meet, not only to other humans but also to members of other species, to fellow inhabitants of the animal kingdom, and the ecosystem of the globe. Precisely because of its radical stance, however, with respect to human–environmental relations, the environmentalist movement tends to fetishize nature, thereby setting it apart from the world of humans. Humans, it is argued, are acting on behalf of nature. The issue of animal rights among radical ecologists "becomes something akin to the activities of the left revolutionaries of the nineteenth century, only now Nature, not the oppressed proletariat, is the beneficiary" (Bennett 1993: 343). Moreover, trapped in objectivist, Western discourse on science and the Other, animal rights activists (oriental environmentalists, if you will) often make a fundamental distinction between "them" (indigenous producers) and "us" (Euro-Americans). In other words, only *some* segments of humanity properly belong to nature, those reported to love animals and take care of their environment, variously called "primitives," the "children of nature," or *Naturvölker*. "We," it is assumed, left "the state of nature" long ago. Similar notions, by the way, have often surfaced in anthropology; thus, deterministic, ecological models are sometimes presumed to apply only to some societies, notably hunting and gathering societies.

Again, an equivalent morality may be revealed in ethnographic practice. In some cases, ethnographers idealize and relativize the world of their hosts, representing their relations in terms of a protective contract. Despite the argument of protection, such a position only maintains the orientalist distinction between the observer and the native. Rosaldo (1986: 96) suggests that the protectionist invocation of "my people" in many ethnographic accounts simply represents an ideological denial of actual relations of hierarchy: "It seems fitting," he claims, with reference to Evans-Pritchard's work on the Nuer, "that a discourse that denies the domination that makes its knowledge possible idealizes, as alter egos, shepherds rather than peasants. Pastoralists, like individual tourists . . . exercise domination less readily than peasants, missionaries, or colonial officials." Similar themes emerge in the academic discourse on textual translation. The idea of the marital contract, as already indicated, is a persistent theme in the works of many literary scholars; thus, the frequent notions of the "fidelity" and "faithfulness" of translation; such constructs even manage to survive the most deconstructive onslaughts. Derrida (1985: 191) speaks of the "translation contract," defined as "hymen or marriage contract with the promise to produce a child whose seed will give rise to history and growth." Johnson (1985: 143) takes the analogy between translation and matrimony into a similar territory, arguing that the translator may be regarded "not as a duteous spouse but as a faithful bigamist, with loyalties split between a native language and a foreign tongue," adding that, perhaps, the project of translation is best described as incest.

Peasants often seem to think of human–environmental relations in terms of protection and reciprocity. Bourdieu gives the impression of a metaphorical extension from the domain of kinship to the sphere of human–environmental relations among Kabyle peasants in Algeria. Kabyle say that the land "settles its scores" and takes revenge for bad treatment and, by extension, the "accomplished peasant 'presents' himself to the land with the stance befitting one man meeting another, face to face, with the attitude of trusting familiarity he would show a respected kinsman" (Bourdieu 1990: 116). Significantly, the relationships between humans and their land are modeled on the social bonds among *distant* relatives characterized by respect and formality, by balanced, not generalized, reciprocity.

In the case of Icelandic fishing, the paradigm of paternalism is represented by the current application of scientific rationality to fisheries management. While fishermen continue to appropriate their prey, in the sense of removing it *from* the natural domain, a world separated from that of humans, with scientific management extraction has been subject to protective measures (*fiskvernd*) and stringent regulations. Consequently, fishermen have become increasingly dominated by techno-scientific knowledge and the agencies of the state. The chief architects of the paternalistic regime of protective fishing and the present system of individual transferable quotas (economists, biologists, and other policy makers) often remain firmly committed to a modernist, objectivist stance. One example is their suppression of the issue of inequality and social distribution, a distracting, ethical subject, an irrelevant externality in the study and management of "economic man," perhaps comparable to the category of "society" in structural linguistics.

Given that, within the moral framework of paternalism, people are aware of the ecological consequence of their actions and that they seek to organize themselves to redress the "balance," the metaphor of the comic plot may seem an appropriate one. The metaphor of comedy has, indeed, been used by several scholars to draw attention to the potential of collective action for corrective environmental purposes. McCay (1995) suggests, for instance, that such a metaphor captures the narrative style of economistic approaches to the question of the commons informed by game theory. She emphasizes, however, that while such approaches represent an important shift in economistic assumptions about human nature, the comic plot is still "squarely modernist" (McCay 1995: 109) in the sense that it fails seriously to address the larger contexts of history, power, and culture. Several anthropologists and economists have raised doubts with respect to the neoclassical and androcentric assumptions of economic theory and the general attempt to separate economics from politics, ethics, and culture (Gudeman 1992, England 1993).

10.4 Communalism

The paradigm of communalism differs from those of orientalism and paternalism in that it rejects the separation of nature and society and the notions of certainty and monologue, emphasizing instead contingency and dialogue. Unlike paternalism, communalism suggests generalized reciprocity, an exchange often metaphorically represented in terms of intimate, personal relationships. The need to develop an "ecological" theory along such lines, a theory that fully integrates human ecology and social theory, abandoning any radical distinction between nature and society, is often recognized nowadays. The outline, however, of such a theory was proposed early on in the writings of the young Marx, who insisted that humans could not be separated from nature, and, conversely, that nature could not be separated from humans. Nature, he argued, "taken abstractly, for itself – nature fixed in isolation from man – is *nothing* for man" (Marx 1959: 169; emphasis in the original).

The development of a theory of practice, including that of Dewey, informed by both the writings of Marx and the perspectives of pragmatism, draws upon these insights. Not only does such a theory provide a perspective that resonates with the paradigm of communalism, dismissing the dualism of experts and laypersons, it also offers a compelling view on how people acquire the skills necessary for managing their lives, starting, as Dewey (1958: 23) put it, "from knowing as a factor in action and undergoing." The theory of practice draws attention to whole persons, master–apprentice relations, and the wider community of practice to which they belong, decentering the study of human action. Such a perspective provides a useful antidote to methodological individualism. The proper unit of analysis is no longer the autonomous individual separated from the social world by the surface of the body, but rather the whole person in action, acting within the contexts of that activity. Similar perspectives have been developed with respect to the notion of the "separative" self in some other disciplines. England (1993) argues that the neoclassical idea of the self and subjective utility – an idea which logically excludes the possibility of interpersonal utility comparisons, of "translating one's own and another person's metric for utility" – must be replaced by the notions of empathy and connectedness.

Recognizing the importance of trust and communalism, anthropologists engage themselves in a serious ethnographic dialogue with the people they visit, forming an intimate rapport or communion. The communalism of fieldwork may be characterized as a project in which anthropologists and their hosts engage in meaningful, reciprocal enterprises, as the inhabitants of a single world. Such a notion has much in common with what Habermas (1990: 85) refers to as the discourse ethics of the "ideal speech situation," a general communicative strategy for recognizing differences and solving conflicts. Once again, there are obvious parallels in literary discourse. Neild (1989: 239) suggests a hermeneutic approach to translation which underlines the reciprocal nature of the enterprise; thus, if the process of translation is to be described as a love affair, an adequate theory of translation must recognize the role of empathy and seduction. The author "reaches out" to the translator, altering his or her consciousness just as the translator alters the text.

Judging from many ethnographies, hunting and gathering societies nicely represent the principles of communalism. In such societies, it is often pointed out, relations with wild animals are characterized by close co-operation. Bird-David (1993) shows how many groups of hunters and gatherers metaphorically extend the communalism of relations among humans to the realm of environmental relations, thereby projecting an image of the "giving environment." Just as a child may expect the care of its parents, the environment provides its *unconditional* support, irrespective of what happened in the past. In hunting and gathering societies, then, human–environmental relations may be described in terms of generalized reciprocity. As the Nayaka of South India say, "forest is as parent." Similarly, the Canadian Cree sometimes speak of themselves as being in communion with nature and animals (Brightman 1993). Hunting activities are frequently regarded as love affairs where hunters and their prey seduce each other; hunters must enter into relationships with game animals in order to have any success and vice versa. To kill an animal is to engage in a dialogue with an inhabitant of the *same* world; animals are social persons and humans are part of nature. In the hunter's view, there is no fundamental distinction between nature and society.

While the classic ethnographic examples, perhaps, of the paradigm of communalism are those of hunters and gatherers, others may be relevant as well. Consider the ancient Scandinavians and their relations to the land. Gurevich (1992: 178) points out that in ancient Scandinavia people were so indissolubly linked with the land they cultivated that they saw in the land an extension of their own nature: "[T]he fact that a man was thus personally linked with his possessions found reflection in a general awareness of the indivisibility of men and the world of nature." Social honor, then, was embodied in the land. Such views are echoed by some Western, academic economists (Nelson 1993: 33) who argue for a "provisioning" definition of economics that considers humans in relation to the world.

To return to the context of fishing, there may be good grounds for exploring, in the spirit of communalism, to what extent the practical knowledge of fishermen could be brought more systematically into the process of resource management and how this knowledge differs from the textual knowledge of professional biologists. I have argued (Chapter 9) that skippers' extensive knowledge of the ecosystem within which they operate, the collective product of apprenticeship, is the result of years of practical enskilment and that it may be wise for management purposes to pay closer attention to this knowledge, allowing for extreme fluctuations in the ecosystem, relaxing at the same time the modernist assumption of predictability associated with the ecological project of sustainability. Some scholars argue that multi-species fisheries are chaotic systems with too many uncertainties for any kind of long-term control (interestingly, in a critical commentary on the idea of "sustainability," focusing on the history of fisheries management, Ludwig, Hilborn, and Walters [1993: 17] make the observation that it may be "more appropriate to think of resources as managing humans than the converse"). But if marine ecosystems are deterministic and chaotic regimes, those who are directly involved in resource-use on a daily basis are likely to have the most reliable information as to what goes on in the system at any particular point in time. In the Icelandic management regime there are few attempts to utilize the knowledge that skippers have achieved during years of practical engagement. There have, however, been some interesting signs of change in this respect, one of which was the so-called "trawling rally," whereby a group of skippers regularly fished along the same, pre-given trawling paths (identified by skippers and biologists), in order to supply detailed ecological information.

It is not quite clear, on the other hand, what the empowering of the practitioner's knowledge entails. While it is true that an extensive body of local knowledge has often been set aside, if not eliminated, in the course of Western expansion and domination and there are good grounds for attempting to recapture and preserve what remains of such knowledge, the reference to the "indigenous" and "traditional" in such contexts tends to reproduce and reinforce the boundaries of the colonial world, much like earlier notions of the "native" and the "primitive"; "natives" and "primitives" have a tendency to congregate in particular times and locations. Where does a particular skill or body of knowledge have to be located to be classified as "indigenous"? How old does it have to be to count as "traditional"? Another contested issue relates to the concept of knowledge itself. Practical knowledge is sometimes presented as a marketable commodity, similar to "cultural capital" – for instance, when encoding indigenous knowledge for the protection of intellectual property rights and defending legal claims about patents and royalties. Much of the practitioner's knowledge, however, is tacit – dispositions acquired in the process of direct engagement with everyday tasks. In reifying practical knowledge we fall into the trap of Cartesian dualism that we may be trying to avoid, separating body and mind.

Given the paradigm of communalism, and the contingent nature of human life, the overly pessimistic plot of tragedy is hardly the appropriate theatrical metaphor for capturing human–environmental relations. Nor is the overly optimistic plot of comedy a convincing one. The members of the human household are not simply greedy Robinsonades (to borrow a Marxian label) who inevitably destroy the ecosystems of which they are a part, nor are they necessarily able to work in harmony for a well-defined common good. The metaphor of romance may be more realistic, allowing for some degree of future hope, in a world with contesting perspectives, conflicting interests, and unexpected turns. In romance, as McCay suggests (following Donham 1990):

> conflict drives the narrative and is not overcome in the manner of neoclassical analyses . . . Romance implies . . . complex development of character, situation, and plot and hinges upon the tension of not knowing what the outcome will be, but hoping for the best. (McCay 1995: 110)

"As a literary metaphor," McCay (1995: 110) concludes, romance "comes closer to the anthropological endeavor."

10.5 Conclusions

I have distinguished three kinds of paradigms with respect to human–environmental relations: orientalism, paternalism, and communalism. Some of the modernist assumptions of orientalism (notably the conjecture of human mastery, the nature/society interface, and the distinction between laypersons and experts) are shared by the paternalistic paradigm – both paradigms are, indeed, the intellectual heirs of the Renaissance, the Enlightenment, and early positivist science (developed by, among others, Descartes and Francis Bacon), all of which instituted a series of decisive dualisms. But while relations of domination characterize the former, protective relations distinguish the latter. Moreover, whereas orientalism suggests the absence of reciprocity in human–environmental relations, the latter typically presupposes human responsibility and balanced reciprocity. Finally, the paradigm of communalism differs from both orientalism and paternalism in that it rejects the notions of certainty and monologue and the radical separation of nature and society. Unlike paternalism, it emphasizes the *generalized* reciprocity of human–environmental relations, an exchange frequently modeled on close, personal relationships. As we have seen, similar relations are evident in ethnographic practice and textual translation. Thus, discourses on environmental management, ethnography, and textual translation have much in common, including the metaphors of personal relatedness and sexual intercourse and the language of theater, the metaphors of irony, tragedy, comedy, and romance.

Social discourse is often, if not always, polyphonic. In modern Iceland, for instance, one can easily elicit evidence for the presence of *all* of the paradigms discussed. To take another example, speaking of Cree representations of human–animal relations Brightman (1993: 194) points out that some indigenous accounts, including the ones of seduction, attest to mutualism and communion in human–animal relations while others indicate hierarchy and domination; such accounts, he claims, can be placed along a "continuum between reciprocity and exploitation." This suggests that paradigms of management should not be regarded as bounded regimes or discursive islands in either time or space. "Operatively speaking," as Dewey (1958: 279) remarked, echoing the Malinowskian idea of the "long conversation," "the remote and the past are 'in' behavior making it what it is." But if Icelanders themselves, or the Cree for that matter, do not seem to be able to make up their minds individually or to agree collectively on crucial ethnographic points – nor, indeed, the ethnographers who have written about them (the issue of "whether Crees believe one or the other model to possess greater validity is exceptionally difficult to address" Brightman [1993: 200] concludes) – how are those with only second-hand ethnography at their disposal to issue a single, final verdict? To this question I can only offer a simple, pragmatic answer: If the problem of ethnographic disagreement needs to be resolved, it has to be approached, much like environmental problems, by means of some form of communicative ethics or a moral standard that allows for free and unrestricted dialogue.

In the early modernist project, with the discovery of the laws of perspective and the triumph of visualism, science became a passionate and aggressive search for truth and knowledge. Later, modernism was exposed as childish and vulgar scientism by critics of various kinds. The project of the Enlightenment was rendered as a metaphysical illusion. Panofsky, who generally emphasized the *successes* of the Renaissance project and its contribution to science, seems to have anticipated some of these developments, suggesting that one may reproach perspective, the "mathematization" of visual space, for "evaporating 'true being' into a mere manifestation of seen things" (Panofsky 1991: 71). Nowadays, Westerners increasingly think of themselves as an integral part of nature as modern environmental discourse seems characterized by a "postmodern condition," a discourse that emphasizes, much like pre-Renaissance thought, the interrelatedness of nature and society, the "individual" nature of human life, in the original, unified sense of the term.

The paradigm of communalism, with its emphasis on practice, reciprocity, and engagement, I suggest, provides an avenue out of the modernist project and current environmental dilemmas. It is true that critics of the modernist project often bask in nostalgia and utopia. The concepts of the perfect society and its antithesis, frequent themes in Western thought, have taken many forms, all of which assume, as Berlin (1989: 120) points out, a Golden Age when "men were innocent, happy, virtuous, peaceful, free, where everything was harmonious" followed by some kind of catastrophe, "the flood, man's first disobedience, original sin, the crime of Prometheus, the discovery of agriculture and metallurgy, primitive accumulation, and the like."

To adopt the dialogic perspective of communalism is not, however, simply to return to the pre-Renaissance medieval world and indulge in naïve romanticism, but rather to embrace a *more* realistic position, shunning the ethnocentric preconceptions of the modernist project. Treating nature, non-human animals, and "other" cultures as mere museum pieces for academic and theoretical consumption is both unrealistic and irresponsible, given the fact that our lives and activities are inevitably situated in larger ecological and historical contexts. Anthropology was led astray by the radical separation of nature and society, what Hollingshead (1940: 358) referred to, in highly modernist terms, as a proper theoretical "beginning."

In the age of postmodernity, Sahlins's (1976: 55) image, referred to above, of anthropology as a prisoner pacing between the "walls" of idealism and materialism, seems increasingly irrelevant. A more appropriate image of contemporary anthropology would be that of a former convict scratching his or her head in the open air, liberated from the Platonian cave, puzzled by the ruins of the prison house – its perceptual illusions, its strict codes of conduct, and its bizarre architectural design. Not only must such ex-prisoners wonder, in Kafkaesque fashion, why they were locked up in the first place and how they eventually got out, but more importantly, how they could possibly enjoy the new freedom in the apparent absence of any kind of idealist agenda but faced with unavoidable materialist constraints and an environmental crisis.

11 Modernity and beyond: the grand aquarium

This chapter focuses on the dualism of nature and society and practical and theoretical implications of anthropological attempts to go beyond it. I shall illustrate my general argument with reference to what I call "the regime of the aquarium," emphasizing the growing role of the state in coastal management and the construction and acquisition of knowledge about marine habitat, particularly in the context of Icelandic fishing. In Iceland, the state has played an increasing role in fisheries management. This development has culminated in a system of individual transferable quotas (ITQs), a highly modernist regime that privileges capital, boat-owners, and scientific expertise, marginalizing labor, crews, and practical knowledge. Anticipated benefits of the ITQ system – in terms of economic efficiency, ecological stewardship, and safety at sea – are less than impressive. More importantly, the system has had far-reaching social implications. For one thing, over time the largest companies have rapidly concentrated quota shares. Also, a semi-feudal system has developed with a fundamental division between quota holders and those who have to rent quota – between "sea lords" and "tenants," to borrow local jargon. A small class of boat-owners has become the de facto owner of the fishing stocks. Similar developments can be observed in several other fisheries (Committee to Review Individual Fishing Quotas 1999). The relative failures of current fisheries management, I suggest, invite a rethinking of the modernist regime and its assumptions of discontinuity, control, and hierarchy.

Aquaria usually owe their construction to the fascination with single species and individual animals. Like keepers of aquaria, marine biologists have typically focused on one species at a time – modeling recruitment, growth rates, and stock sizes – although recently they have paid increasing attention to analyses of interactions in multi-species fisheries. Fisheries management seeks to systematically affect the structure of fish populations by controlling the relative sizes of different species and year classes, through regulations concerning fishing effort, gear, mesh sizes, territorial restrictions, and so on. They may be difficult to administer at times and the results are not necessarily along the lines envisaged by those in charge, but it is generally assumed that things are "under control." The dominant management response to the current problem of overfishing in Western countries, including Iceland, is characterized by a preoccupation with single species, certainty, and expert control. That response, I suggest, is part and parcel of the dualistic, modernist project. Not only does it underline the boundary between the inside and the outside, observers and the observed, it fails to appreciate the nature and role of practical knowledge, misconstruing the relationships between humans and the environment.

The regime of the aquarium is not restricted to fisheries; as Scott (1998) has shown it goes with state power in high modernity. Indeed, some of the properties of that regime were laid out early on in Western history. Foucault (1973: xi) has argued in his work on the "birth" of the medical clinic that modern medicine signified a fundamental shift in the understanding of the human body, a "mutation in discourse"; in particular, the body was dissected and subjected to the scientific gaze and manipulation of supposedly detached observers. Western understanding of the ocean and its inhabitants has undergone a similar transformation. It is pertinent, paraphrasing Foucault, to speak of the birth of the aquarium. The oceans tend to be seen as a gigantic fish tank, scientifically managed for human purposes. Such a notion is the culmination of a complex cognitive and political history. Until fairly recently Westerners typically assumed that the supply of living resources in the ocean was a boundless one. Such a position, of course, was untenable in the long run. Many of the world's major fishing stocks are threatened with overfishing, global warming, and pollution – oil, radioactive waste, and other byproducts of human activities – and fisheries more and more resemble other branches of industries in that the resource base is increasingly subject to deliberate human impact. For one thing, the boundaries of "wild" fisheries are increasingly becoming blurred, with exponential growth in sea ranching and fish farming, not to mention genetic mixing and engineering. Consequently, to think of the oceans as a boundless storehouse of living resources unaffected by humans really does not make much sense any longer. It would be far more appropriate to speak of the regime of the aquarium.

11.1 Environmental anthropology

The concept of modernism usually connotes at least three related characteristics: the dualism of nature and society, the notion of objective science, and the assumption of linear control. Thus Gudeman (1992: 151) defines the "modernist production regime" as a regime based on the idea "that the human and natural world can be organized and subjected to rational, totalizing control." Scott (1998) uses similar terms. For him, "high modernism" is represented by

> supreme self-confidence about continued linear progress, the development of scientific and technical knowledge, the expansion of production, the rational design of social

order, the growing satisfaction of human needs, and, not least, an increased control over nature (including human nature) commensurate with scientific understanding of natural laws. (Scott 1998: 89–90)

Nature, then, is presented as an inherently logical and linear domain and, accordingly, the project of the resource manager is likened to that of the engineer or the technician. The modernist approach to environmental problems, with its separation of nature and society, draws upon textual notions of scientific practice developed during the Middle Ages when it was customary to speak of nature as "God's book" and to regard science as the "reading of the book of nature." Before the advent of modernism there was no radical separation of nature and society in European thought; people saw themselves as integral parts of the world, embedded in nature. In a brief period, nature became a quantifiable, three-dimensional universe appropriated by humans. This universe represented a radical departure from the earlier, enclosed universe of the Aristotelians constituted by the Earth and its seven surrounding spheres.

The distinction between nature and society is central to both modern science and modernist culture. Such a dualism has not only been reinforced by a rigid academic division of labor and massive institutional structures, it also tends to be "engraved" in the financial and spatial organization of universities and campuses, in their architecture, layout, and budgets. Much social scientific thinking shares the main assumptions of the modernist perspective. Interestingly, the prologue to M.C. Bateson's (1984) memoir of her parents, G. Bateson and M. Mead, is subtitled "The Aquarium and the Globe." Apparently, her parents used globes and aquaria to give their daughter a sense of the integrity of the biosphere and the necessity of "building" a rational world for future generations, "to balance the needs of living creatures and their relationships with each other, the cycles of growth and respiration and decay" (Bateson 1984: 5). As metaphors, both globes and aquaria are highly modernist constructs; most importantly they position the observer outside the system observed, gazing at a separate reality, much like the medical practitioner in the Foucaultian clinic.

If disembeddedness, dualism, certainty, and human mastery are the characteristics of modernism, *post*modernism suggests the opposite – namely, embeddedness, monism, and the absence of certainty and human mastery. For me, postmodern social science is committed to continuity, engagement, and the negation of the notion of the detached observer. It seems that at the dawn of a new millennium anthropological theory on human–environmental relations is rapidly moving in such a direction, resounding the *condition* of postmodernity (Harvey 1989), the declining faith in rigid dualisms, hierarchy, and the tendency to conflate theoretical spaces previously kept separate. In recent years, indeed, the distinction itself between nature and society has increasingly been subject to critical discussion in several fields, including anthropology. And there are good grounds for second thought. For one thing, humanity has an embodied physicality that, by definition, naturalizes it, as Marx maintained long ago. Also, modern humans are presented with a "nature" very different from that experienced by earlier generations. Biotechnology and genetic engineering have revolutionized our capacity to analyze and alter DNA material, raising new and fundamental questions as to what constitutes "life" and "nature"; organisms are engineered and manufactured according to human designs and for human purposes. Human nature, then, must be a fleeting category.

Moreover, recent theoretical developments in biology have questioned the classic argument of Mendel and Darwin that organisms are autonomous objects dictated by genes and selective pressures. An emerging, alternative model emphasizes that the organism is empowered to shape its own development, the *subject* of evolutionary forces. The dialogic vocabulary of "co-evolution" and "niche construction" (Odling-Smee 1994) seems to be emerging in place of mechanical Newtonian notions. Any distinction between inside and outside (and, by extension, between nature and society) seems beside the point. It seems reasonable to assume that humans are simultaneously part of nature and society and that modern policy on the environment should be based on *that* premise, and not on the idea that humanity, or some part of it, is suspended above nature.

While criticism of the modernist project and its separation of nature and society has been fueled by recent developments, including the greening of public discourse, it is not a brand new phenomenon. The sixties and seventies, in particular, were characterized by growing doubts and discontent. Bateson, for instance, offered an early warning to the nature/society divide and the idea of absolute human domination, emphasizing that humans are part of nature, not external "autocrats." The "arrogant" Western notion of "complete power over the universe," he argued, was obsolete; in its place there was "the discovery *that man is only a part of larger systems* and that the part can never control the whole . . . *he cannot have a simple lineal control* . . . Life is not like that" (Bateson 1972: 413; emphasis added). Vayda and McCay (1975) challenged the cybernetic notions of homeostasis and explanation in the so-called "new ecology" of the time. And Wagner argued, from a different, "semiotic" perspective, that the distinction between the environment and the environed was an arbitrary one, fixed for the sake of analysis. It is worth quoting at some length an important and somewhat neglected paper:

> Positivistic epistemology has generally favoured the notion of "levels" in the sense that the cultural is said to be an "abstraction from" nature – a replication of its "orders" via human artifice . . . In fact . . . the distinction is itself non-locatable, . . . nature is as much abstraction from culture as the cultural is an abstraction from nature. (Wagner 1977: 395–396)

Wagner's conclusion has a clear postmodern twist. In his view, nature and semiotics are so "completely and mutually continent of one another . . . that no boundary of any sort can be established between them"; this implies, he argues, "that the Cartesian duality is at once completely insoluble and largely irrelevant" (Wagner 1977: 409).

Postmodern criticism has undoubtedly enriched anthropological theorizing on environmental issues, drawing attention to relations of power, to Western anthropocentrism, to the problems with dualism, and to the inadequacies of the correspondence or mirror-of-nature theory of truth and the grand narratives of modern science on progress and control. Some postmodernist thinkers, however, leave it unclear how people should act with respect to the environmental crisis, dismissing the environmental problem as a "social construction." As Gare (1995: 97) argues, while the idea of a "global environmental crisis" can be shown to serve those who are attempting to mobilize people to address it, with their hostility to grand narratives radical postmodernists "leave environmentalists no way to defend their belief that there is a global crisis or to work out what kind of response is required to meet it . . . They are bound by assumptions which make the idea of a global environmental crisis incomprehensible." One wonders, indeed, given the social-construction perspective, how humans could possibly stumble on solutions to their problems.

Recognizing the contributions of postmodernist approaches to environmental issues, I suggest they also may have important drawbacks. I fail to see how one can develop an effective politics of the environment without some grand narrative, in the absence of any kind of theoretical authority in which we could ground reasonable and responsible claims about the nature and scale of the environmental problems we face. The Earth is a place to dwell in and to maintain its integrity and to avoid ecological bankruptcy we have to at the same time dwell and attempt to "manage." Grand narratives, whether we like it or not, seem to be a political and environmental necessity. Such narratives, however, should be constructed through a democratic process and, moreover, they should combine theoretical expertise and practical knowledge, the "cunning intelligence" summarized in the Greek term *mêtis* (Scott 1998). And that brings me to my empirical example, modernist approaches to fisheries and their problems.

11.2 Icelandic fishing

Icelandic production discourse has undergone a series of successive changes as Icelanders have assumed new kinds of relations in the course of appropriating marine resources. To each phase in the development of Icelandic society corresponded a particular dominant "paradigm," an underlying framework of understandings and assumptions with respect to what constitutes production and ecological expertise. One of the important changes concerns the discursive shift from land to sea at the turn of the last century. During earlier times, Icelandic farmers and landowners occupied a central position and, consequently, fishing was regarded as merely a supplementary subsistence activity. Fishing was not just a marginal occupation; it was also the subject of a cultural struggle. This can be seen from the fact that in the nineteenth and early twentieth century those in power tended to present fishing communities as "devoid of culture" (*menningarsnauð*), the source of degeneration, alienation, and deficient language. In the nineteenth century, however, as new markets for Icelandic fish were developed, especially in Spain and England, fishing villages grew and there emerged an expanding market economy. While for many Icelanders the agricultural community continued to provide the dominant cultural framework, the "essence" of the Icelandic way of life, the focus of discussions on economics and production inevitably shifted from the landed elite to the grass roots of the fishing communities as fishing became a full-time occupation and a separate economic activity. Early in the twentieth century, fishers and boat-owners gradually became the central agents of production discourse, replacing the landed elite as the economy shifted from stagnant agriculture to expansive fishing. In the process, agriculture was redefined as a burden to the national economy. Now, once again, with scientific management and a quota system in the fisheries, the discursive pendulum has swung in the opposite direction – from sea to land. Fishing remains a major economic enterprise, but the makers of knowledge and economic value are no longer fishers but the land-based owners of boats and fishing plants and the holders of scientific, textual knowledge.

With Iceland's independence in 1944, fishing effort on Icelandic fishing grounds multiplied. In 1948 the Icelandic Parliament passed laws about the "scientific protection of the fishing grounds in the coastal zone," in order to prevent overfishing of its major fishing stocks, particularly cod. Four years later, Iceland announced that it would extend its territorial jurisdiction from 3 to 4 miles and in 1958 it unilaterally extended its jurisdiction to 12 miles. In 1976, the Icelandic government extended the national fishing limits to 200 miles which marked the end of the last Cod War with Britain and West Germany. The domestic fishing fleet, however, continued to grow and catches, relative to effort, continued to decline. The first serious limitations on the fishing effort of Icelandic boats were temporary bans on fishing on particular grounds. By 1982 politicians and interest groups were increasingly of the opinion that more radical measures would be needed to limit effort and prevent the "collapse" of the cod stock. A boat-quota system was suggested in 1983 to deal with the ecological and economic problems of the fisheries, a system that would divide this reduced catch within the industry itself. The precise allocation of catches was debated, until it was agreed late in 1983 that each boat was to be allocated an annual quota on the basis of its average catch over the past three years. This meant that some boats would get higher quotas than the rest of the fleet, a fundamental departure from traditional policy. And quotas were allocated to boat-owners, not crews. These developments highlight two important points, as the following discussion will show: the relative marginalization of the practical knowledge of fishers and the shift in the locus of power from labor to capital.

11.3 Experts and laypersons

In Iceland, some marine biological research already occurred by the end of the nineteenth century, but full-time research started later, in the 1940s. At the beginning of the twentieth century, most of the contributions to *Ægir*, the journal of the national Fisheries Association (which embraces most interest groups), were those of fishers, but gradually marine biologists entered the scene. With fishers being important agents in the expanding economy, marine scientists had to carve a space for themselves in the role of collaborators and apprentices. This is especially clear in the writings of Bjarni Sæmundsson, the pioneering ichthyologist. In the 1890s he traveled throughout Icelandic fishing communities to learn from practicing fishers:

> I had the opportunity to observe various kinds of newly-caught fish, to look at fishing gear and boats and to listen to the views of fishermen on various matters relating to fishing and the ... behaviour of fish. (*Ægir* 1921: 115)

Sæmundsson seems to have thought of himself as a "mediator" (*milliliður*) between foreign scientists and Icelandic fishers (*Ægir* 1921: 116), eager to learn from both groups. Scientific knowledge, along with the "practical knowledge" (*reynsluþekking*) of fishers, he suggested, was "the best foundation for ... the future marine biology of Iceland" (*Ægir* 1928: 102).

The pioneering biologists did not only regard themselves as humble apprentices, they were moderately optimistic about the immediate achievements of the scientific enterprise. Referring to the prospects of dealing with "the old mystery, the migration of fish," Sæmundsson comments:

> We should not expect ... to be able to deal with everything and, thus, to answer whatever question we may have, for instance to establish the location of herring ... at a particular time. That kind of knowledge is far away, although it is our mission (*hugsjón*) to be able to provide it in the future. (*Ægir* 1924: 144)

Such a mixture of mission and modesty was, no doubt, necessary in the beginning, to provide political and financial support for marine science. Gradually, the necessary trust and confidence were attained; fishers, boat-owners, and the general public participated in establishing fisheries science. Significantly, in 1931 one of the regional fisheries associations resolved:

> The behaviour of most of the fish species we exploit is now known in most respects. The place and time of spawning ... are topics which science has for the most part mastered. Nevertheless, there are many "dead spots" in our knowledge of fish behaviour. Little or nothing is known about what determines fluctuations in the catch. (*Ægir* 1931: 29)

At the same time, however, another discourse emerged which downgraded the expertise of fishers. While the editor of the fisheries journal urged fishermen to participate in discussions on the fisheries, he sometimes reinvented the biases of the earlier agricultural elite:

> You fishermen! This journal is intended for you, it should be your guide and your voice ... it should speak for you when you are busy at sea ... it should enlighten those of you who live at the outskirts, at the margin where the profit is, but often, too, unfortunately, ignorance and poverty. (*Ægir* 1932: 159)

By the middle of the century, the subtle competition of the discourses of fishers and biologists seems to have developed into open confrontation. In 1947 the fisheries journal published an article that found it necessary to remind the readers of the journal of the significance of practical knowledge:

> Fisheries research is ... intimately connected to fishing ... and naturally ... it should be carried out in collaboration with perceptive fishermen and boat-owners ... *This may not be particularly scientific, but we should keep in mind that the experience that perceptive fishermen have acquired after years of practice ... must provide some kind of guidance to the scientists. It is no coincidence that the same men catch more than others year after year.* What matters most is their attentiveness and their perceptiveness with respect to the behaviour of cod and herring. Icelandic ichthyologists should recall Bjarni Sæmundsson who once remarked, in his well-known humble spirit, that he owed most to the fishermen of this country (*Ægir* 1947: 159; emphasis added)

During the cod wars, the biologists at the Marine Research Institute, established in 1965, emphasized the prospects of estimating the composition and size of the cod stock:

> In recent years, the success of spawning for a given year has been extensively studied. We have obtained tentative estimates of the size of different year classes, but since such research only began recently ... it is not quite clear ... what each year class will supply for the fisheries. Before long, however, we should know, and then we expect to be able to predict, only a few months after spawning, the real size of the year classes. I am optimistic that in the future we will be able to make forecasts with more accuracy than at present. (Schopka 1975: 48)

Now, decades later, these words sound overly optimistic. Fundamental ecological relationships – including the relationships between the size of the spawning stock, the success of spawning, the size of the future fishing stock, and fishing effort – have turned out to be far more complex and difficult to establish than the biologists estimated.

The tone of humility and mutual learning typical for the pioneering biologists during the first half of the century has been replaced by claims about scientific certainty and folk "misunderstanding" (see *Ægir* 1964: 109). The element of trust that characterized relations between scientists and fishermen evaporated. One skipper provided the following observation:

> When fisheries biologists realize that fishermen possess knowledge that they themselves do not have, and when these two groups begin to co-operate on the basis of each other's knowledge, then we may envisage realistic knowledge about the quantity and behaviour of cod on Icelandic grounds. (Guðjón Kristjánsson, *Ægir* 1979: 595)

Another skipper remarked a few years later: "Fisheries biologists have no possibility of finding or counting the fish in the sea. They have no equipment for this purpose beyond those that fishermen have" (*Ægir* 1986: 33).

Skippers' approach to the environment and the acquisition of ecological knowledge is very different from that of marine biology. Formal schooling is essential for Icelandic skippers, but they all seem to agree that most of their learning takes place "outdoor," in the course of fishing. Skippers' extensive knowledge of the ecosystem is the result of years of practical enskilment, the collective product of a community of practice. Skippers discuss their own research strategy as a dynamic and holistic one, allowing for flexibility in time and space. Usually, their accounts emphasize constant experimentation in the flux and momentum of fishing, the role of perpetual engagement, and the importance of "hunches," intuition, and tacit knowledge. Somewhat ironically, the skippers' approach is much closer to the postmodernist paradigm than the approach of marine biologists.

The practical knowledge of skippers and their crews, of course, is just as essential for the success of fishing expeditions as before. As we have seen, however, in a few decades fishers' knowledge has increasingly been marginalized in public discussions and policy making. Why did this shift in discourse take place? One reason may have to do with changing paradigms in marine biological research. More importantly, though, Icelandic high modernity gradually ensured a strong connection between the state, capital interests (the owners of vessels), and marine science. During the Cod Wars of the 1970s, marine science played an important role; scientists supplied the arguments necessary for the "rational" management of the fisheries. In the words of Jón Jónsson, former Director of the Fisheries Research Institute: "Our rights will never be accepted internationally unless we are able to supply solid evidence, based on scientific research, for the dangers posed to our fisheries and our whole cultural existence" (*Ægir* 1947: 258). With the birth of the modernist fisheries regime and the grand narrative of marine science, the voice of fishermen was gradually subdued, if not silenced.

While the present management regime represents the apex of modernist management, with science in the leading role, there is a growing recognition of the rhetorical context and the political role of marine science. This was triggered by discontent with the prognoses of the Marine Research Institute and its conservative recommendations regarding the total allowable catch. At the same time it illustrates the thin line between marine science and politics. The results and recommendations of both marine scientists and the economists responsible for the establishment and design of the ITQ system have also been challenged from within the scientific community, notably by researchers at the University of Iceland.

Since the introduction of ITQ management in 1984, and particularly in the years following the enactment of the fisheries legislation of 1990 whereby fishing rights became true commodities (fully transferable), the Icelandic fishing industry has undergone a radical transformation. ITQs have become increasingly concentrated in the hands of large vertically integrated companies, while the number of smaller operators has diminished (see Palsson and Helgason 1995 for a detailed early account of these structural changes). A large majority of participants in the fishing industry and the Icelandic public are deeply concerned about the concentration of fishing rights in the hands of a few. In public discourse the owners of the biggest companies are habitually referred to as "quota-kings" (*kvótakóngar*) and "lords of the sea" (*sægreifar*). Complaints are often raised that while fishing rights were traditionally the birthright of all Icelanders, now that they have become commoditized they will be inherited by the holders' descendants like any other privately owned item.

Not only are quotas being concentrated with the biggest companies, new relations of "tenancy" have developed. These typically involve long-term contracts between large ITQ-holders and smaller operators, where the former provide the latter with ITQs in return for the catch and a proportion of the proceeds. One such arrangement, usually referred to as "fishing for others" (*veiða fyrir aðra*), has become increasingly widespread. In such transactions, the supplier of the ITQs is a large vertically integrated company that controls a processing plant and a trawler. A contract is arranged whereby the large ITQ-holder transfers ITQs to the smaller operator's boat. The latter then fishes the ITQs and delivers the catch to the suppliers' processing plant for a price way below the free market price. Understandably, the lessee boat-owners cannot make the same level of profits when fishing for others in comparison to fishing their own ITQs. As a result, they try to compensate for their losses by reducing the wages of fishers. Another form of ITQ-leasing, generally referred to as "ton-for-ton fishing" (*tonn á móti tonni*) takes place when a large ITQ-holder offers to pool one ton of his ITQs against every ton put forward by the lessee. The latter then goes out and catches the fish and delivers the catch to the lessor's processing plant. In some cases of ton-for-ton fishing, regional fish markets act as the ITQ-brokers. Like the processing plants the fish markets require a steady supply of raw material, and in order to achieve this they sometimes buy a small "surrogate" fishing boat on which to "store" their ITQs (to hold ITQs an operator must control a fishing vessel). These surrogate boats are rarely used for fishing; in fact, they are often unseaworthy.

The typical lessee operator is an owner of a relatively small vessel with a meager supply of ITQs or the owner of a so-called "eunuch" (*geldingur*) – a boat that has virtually no ITQs of its own and is solely operated on leased ITQs. Through ITQ-leasing boat-owners with small ITQ-holdings manage to prolong their fishing operations throughout the year. For the lessors of ITQs, however, participation in these new relations of production represents a rather lucrative business. By leasing its ITQs, a company can free itself from the expenses of actually catching the fish, while still procuring up to half of the market value of the resulting catch. These developments have led fishers to augment existing feudal metaphors by referring to the ITQ system as a tenancy system (*lénskerfi*). In this conception the so-called quota-kings are likened to medieval landlords, with the fishing grounds as their estate (*óðal*). Conversely, fishers and small-scale lessee operators become the "tenants" or "serfs" (*leiguliðar*), who are granted access to the fishing grounds on the prerequisite that they hand over their catch to the lessor's processing plants. The lessor not only controls how many tons a tenant boat is allowed to fish, but also the duration and location of each fishing trip. As one skipper put it, "one must give in to almost every demand, because the quota-king makes all the rules, sets the price and everything."

The discourse of resistance briefly described above seems to highlight the issues of agency and a labor theory of value (Helgason and Palsson 1997). In the case of the Icelandic ITQ system, the allocation of commoditized fishing rights to boat-owners has resulted in a privileging of capital over labor, shifting power from sea to land and widening the economic rift between boat-owners and their employees. Commenting on a young boat-owner who had given up fishing and turned to renting out his ITQs, one fisher remarked, "he isn't working and that's unnatural, a young man like him!" In the words of another fisher, "it necessarily adds a devilish aspect to the system when people can rent their quota and then just relax in bed." These and related statements testify to a powerful folk concept of work, bodily experience, and labor value that resonates with the so-called medieval house view of production and exchange; wherein the merchant and the usurer were not held to be creating anything but immorally breeding money from money.

The issue of the ownership of fish and quotas is frequently contested. ITQs remain, according to the first clause of the 1990 fisheries management legislation, the "public property of the nation." During debates on the 1990 fisheries laws, some members of the Icelandic Parliament raised doubts about the "legality" of the ITQ program, arguing that proposed privileges of access might imply permanent, private ownership that contradicted some of the basic tenets of Icelandic law regarding public access to resources. The laws that eventually were passed categorically stated that the aim was *not* to establish private ownership. The real world of legal and economic practice, however, seems to have a momentum of its own. While quotas, according to the law, are not to be regarded as the private property of quota holders, quota shares may achieve the characteristics of private property as time passes. There has been a long discussion over whether quotas can be used as collateral for obtaining loans. The law seems unclear on this point, but economic and legal practice seems increasingly to recognize quotas as collateral, which is a further step in the recognition of quotas as private property, undermining the significance and effect of the statement in the current law on public ownership. Some evidence indicates that quota shares are gradually acquiring the characteristics of full-blown private property.

11.4 Beyond the modernist aquarium

The currently fashionable approach to fisheries in many Western countries, typically based on the allocation of individual property rights to fishing stocks in terms of shares in the "total allowable catch" (TAC) for a given species, has been developed in several contexts beside Iceland, including Canada, New Zealand, South Africa, and the United States. By instituting private property rights to the fishing stocks in the form of quotas, and letting the market regulate their distribution, it is claimed, rational production will theoretically be attained. Assuming a sense of responsibility among the new "owners" of the resource (the quota holders) and an unhindered transfer of quotas from less to more efficient producers, it is argued, privatization both encourages ecological stewardship and ensures maximum productive efficiency. The regime of fisheries management discussed above is informative in this context. The Icelandic ITQ system represents one of the pioneering ITQ systems in the world and as a result, perhaps, it reveals both the explicit and implicit assumptions of the theory of ITQs and the modernist perspective, with its emphasis on control and dualism. Also, in Iceland it may be relatively easy to detect and observe both intended and unintended consequences of modernist fisheries, since here, unlike many other contexts, including those where ITQs have been practiced, fishing is of central importance to the national economy. The ITQ "experiment" of social, economic, and ecological engineering, in a sense, is relatively uncontaminated by confounding factors.

As we have seen, the Icelandic case illustrates many of the pitfalls of modernism, the radical separation of nature and society and the preoccupation with order, hierarchy, scientific privilege, and control. Here a totalizing regime that concentrates wealth and power in the hands of a few boat-owners and marginalizes the practical knowledge of fishers and other "laypersons" has been established under the banner of progress and modernist science. Not only is this regime grossly unfair, engendering inequality on an unprecedented scale, there are good grounds for questioning its reported beneficial role for sustaining fishing stocks. First, in attempting to remove both the economic actor (the quota holder) and the observer (typically the economist) from the realm of social relations and communitarian concerns, implementing the autonomous homo economicus of neoclassical economics and theorizing of "tragedies of the commons," ITQs tend to manufacture irresponsible resource-users. Where ITQs have been instituted, cheating (the underreporting of catch) and the dumping of bycatch (low-value species and fish for which the producer in question has no quota) are identified as major problems of "monitoring."

Secondly, given the chaotic processes of marine habitat, there is no guarantee that the practices of allocating total allowable catch and individual transferable quotas have the consequences anticipated by theorists and policy makers. Many environmental "goods," it may be argued, need to be understood in a holistic manner, not as atomistic objects but as parts of a larger complex. Acheson and Wilson (1996) suggest that the "numerical" approach of current resource economics and marine biology, an approach that has much in common with the regime of the aquarium, emphasizing single species, linear relationships, and states of equilibrium, fails to account for the realities of many fisheries. Their empirical work shows that while fisheries are deterministic systems, because of their extreme sensitivity to initial conditions ("butterfly effects," in the language of chaos theory) even simple fish communities have no equilibrium tendency. As a result, management faces forbidding problems when trying to explain the noise in ecological relationships. For example, it has been said about the relationship between recruitment and stock size, often a key issue for managers, that the degree of accuracy required for prediction is beyond any capabilities we might expect to achieve in a fisheries environment. Therefore, it becomes difficult, if not impossible, to know the outcomes of management actions such as quotas. This partly accounts for the failures of many attempts to manage fisheries, although the sheer level of fishing effort is, no doubt, a major problem generally.

The relative failure of the modernist regime and individual quotas of recent years to deliver the goods they promised and the social repercussions they entail suggest that it may be wise to look for alternative management schemes and alternative ways of understanding human engagement with nature. Pragmatism, with its emphasis on phenomenological perspectives, practical engagement, and direct perception, offers a way out of the modernist regime. According to the pragmatist school, the practitioner's knowledge is an emergent phenomenon, situated in immediate experience and direct engagement with everyday tasks. Such a view, which starts with the assumption that all behavior is the result of the collaboration of person and context and that "personal" capacities arise from the mutual relations of individual and environment, suggests a radical break with the Cartesian tradition, the detachment of the subject. A similar perspective may be applied to scientific practice. Thus, the idea of some Archimedean standpoint outside nature and history is frequently subject to critical discussion on empirical, ethnographic grounds, with the growing awareness that the modernist perspective fails to account for the actual practice of modern science. The proper focus should not, therefore, be the passive autonomous individual, but the whole person acting within a particular context. The Cartesian project has also been challenged by another related development: by recent theorizing on the human body that emphasizes the embodied grounding of cognition, experience, and learning. Much of our knowledge, according to this perspective, is tacit, dispositions engraved in the *habitus* of the mindful body.

The denigration of practical knowledge, evident in the Icelandic case, is a byproduct of high modernism. Scott (1998: 305) argues there are three reasons for this: the "professional" concern that the more the practitioner knows the less the relative value of the expert, the general modernist contempt for everything doing with the past, and, finally, the fact that practical knowledge tends to be codified and represented in a form which is alien to science. While *mêtis*, the intuitive intelligence of the practitioner and the painstaking attention to detail and the demands of here and now, is indispensable for the success of any kind of practical enterprise, it tends to be discursively relegated to the margin; high modernism, Scott (1998: 331) suggests, "has needed this 'other,' this dark twin, in order to rhetorically present itself as the antidote to backwardness." The proper response to the modernist agenda is not romantic adherence to the past, the fetishing of "traditional knowledge," but rather a management framework which is democratic enough to allow for a meaningful dialogue between experts and practitioners and flexible enough to allow for a realistic adaptation to the complexities and contingencies of the world – in sum a communitarian ethic of "muddling through." Those who are directly involved in resource-use on a daily basis may, after all, have highly valuable information as to what goes on in the sea at any particular point in time. It is important to pay attention to the practical knowledge of skippers, allowing for contingency and extreme fluctuations in the ecosystem. Some form of self-governance in fishing may be a practical necessity.

Self-governance, however, on the basis of practitioners' knowledge, may invite a dubious commodity fiction of its own. In orthodox theories of learning, knowledge often becomes analogous to grammar or dictionaries. Given such a perspective, indigenous knowledge is sometimes presented as being like "cultural capital." Much of the practitioner's knowledge is tacit – dispositions inscribed in the body in the process of direct engagement with everyday tasks. A thorough discussion of what constitutes tacit knowledge and how it is acquired and used seems essential for both re-negotiating the hegemony of scientific expertise and rethinking the relationships between humans and their environment. In this process, anthropologists can have a crucial role to play, given their ethnographic method and their routine immersion in the reality of the practitioners.

No doubt, the dualism of nature and society was highly useful in the sense that it made room for social scientific approaches to the environment within academies and universities traditionally dominated by natural science. As Benton (1991: 25) argues, "dualistic, anti-naturalistic programmes in the contemporary human sciences are best understood as primarily defensive reactions to the intellectual imperialism (and, in many cases, moral and political conservatism) of biological reductionist programmes." Just as the dualism of modern social science was necessary to draw attention to the "collective" aspects of social life, to confront the individualistic bias of psychology, a dualistic academe with a relatively autonomous wing for social science was necessary to facilitate systematic discussion of some of the "social" aspects of the environment previously neglected by the paradigms of natural science. We may wonder, therefore, what will be the consequences if the nature/society dualism is suppressed. Will it mean that the projects of "development" and the "environment" become, once again, subject to technological fetishism and green revolutions, relegated to biology, genetics, and engineering – with all the (im)practical implications such reductionisms have had in the past? Is it necessary, perhaps, given potentially shrinking budgets for social scientific research and the power struggle within Western universities and bureaucracies, to staunchly defend a well-demarcated social scientific camp? Surely, budgets are important, but I don't think that stubborn adherence to the dualisms of the past is a realistic defense strategy, to address current problems of funding. Social scientists, in fact, might have been more successful lobbyists had they rejected the duality of the individual and the collective, the natural and the social.

11.5 Conclusions

There are profound problems, as we have seen, with modernist approaches to the environment. Icelandic fisheries management demonstrates many of the results. The system of individual transferable quotas, with its radical separation of nature and society and its division of winners and losers, both manufactures environmental irresponsibility and privileges capital and scientific expertise, marginalizing labor and practical knowledge. To move beyond the modernist approach represented by what I have termed the regime of the aquarium, it is essential to rethink human–environmental relations, including the nature/society divide. I have argued for a fundamentally revised division of academic labor; in particular the removal of the disciplinary boundaries between the natural and the social sciences. Such a move should not be seen to violate the integrity of the discipline of anthropology; on the contrary, from early on anthropology refused to categorize as either natural or social. Anthropological practice, in my view, should be broadened, emphasizing intensive collaboration with a variety of "other" disciplines touching on environmental issues. The anthropology of the environment should be a dialogic affair.

Ellen has suggested that "[t]he more you talk about nature the more you create a meta discourse which relies upon its existence, and the more you give it a life as a ruling concept; in trying to get *beyond* nature and culture we reify an opposition" (1996: 29; emphasis in the original). That meta discourse, of course, as Ellen implies, is not an end in itself. However, it has been around for quite a while, and silencing it is unlikely to dissolve the opposition of nature and society; dualisms just don't disappear because people stop talking about them. More crucially, there is an important sense in which this critical discourse *is* a highly useful one; it provides a framework for identifying and removing persistent obstacles to the understanding of human engagement with the environment. An important item on the management agenda is to understand practical ecological knowledge and the ways in which it might be brought more efficiently than at present into the process of resource management. Given the perspective of practice theory, practical expertise is the result of a simultaneous engagement of the human actor with fellow humans and the rest of the material world – to the extent that the distinction between the two "environments" of nature and society has no meaning.

Why, one may wonder, has the long theoretical conversation that sees humans *in* nature, engaged in situated, practical acts, been subdued most of the time in Western thought? While the dualism of nature and society has a dynamic of its own, driven by industrial capitalism and the successes of modern science, monism, too, has its critics. Plumwood (1991), writing on gender and ecology, suggests that one cannot really care for the environment if it is "simply" an unbroken extension of oneself. Such a position echoes the developmental views of Grimshaw (1986) with regard to dependent selves, the argument that too much conflation in a dyadic relationship (in marriage or a parent–child relationship, for instance) may have detrimental effects for the personal development of one or both of the partners – namely, a loss of autonomy or maintenance of dependency. Grimshaw (1986: 182–183) suggests that care and understanding "require the sort of distance that is needed in order not to see the other as a projection of self, or self as a continuation of the other." For Plumwood (1991: 14) these points "seem to . . . apply to caring for other species and for the natural world as much as they do to caring for our own species." While the loss of self-boundaries may hinder personal growth, resulting in neglect rather than care, Plumwood's environmental argument is not convincing. After all, the extended notion of the embodied self – of *being* a body and not simply having it – does not preclude the idea of bodily concern. Indeed, caring for one's health and fitness is a major, if not obsessive, preoccupation nowadays, and not simply among those who project their body as a fetish external to themselves. The dissolution of the mind/body dualism, currently fashionable among Euro-Americans, is frequently associated with healthy diets and care for the body. And if, for many people, the incorporation of the body into their notion of self signifies intensive bodily care, why should they neglect the environment once they reject the dualism of nature and society?

I have found it useful to draw upon Scott's (1998) treatise about high modernism. Scott's work tends to focus on agrarian discourse and the metaphor of gardening, a metaphor underlining the reconstruction of natural sites and the designed terrestrial space of botanical order, although it also refers to fishing. As a metaphor, the aquarium is no less compelling than the botanical garden. I hope to have shown that the dualism of nature and society, one of the corner stones of the modernist perspective and the regime of the aquarium, is *part of* the environmental problem in that it both obstructs understanding of the human predicament and the kind of change in human–environmental relations which is needed. It is important to develop an approach that fully integrates human ecology and social theory, adopting a monist perspective that conflates nature and society. Given such a perspective, the aquarium may be an effective key metaphor for the interconnectedness of the biosphere. For such a metaphorical association to make sense, however, one would have to relax some of the modernist assumptions. The aquarium will have to include the practitioner of science as well as the "layperson," swimming – along with fish and other earthlings – in the household of life. For me, that amounts to a postmodernist anthropology of the environment, an anthropology sensitive to interconnections, relations of power, and social discourse.

12 Housekeeping: *Oikos* and the Anthropocene

It is man's earth now. One wonders what obligations may accompany this infinite possession.

Henry Fairfield Osborn Jr., 1948[1]

This chapter discusses the notion of the Anthropocene and its implications for anthropology and environmental discourse, exploring the usefulness of notions of the *Oikos* and "house economies," documented in many historical and ethnographic accounts, for the challenges of the Anthropocene. Famously, anthropologist Roy Rappaport (1967), one of the pioneers of "ecological" anthropology, carefully explored relations among humans and pig populations in New Guinea, their representations in local discourse, and their ritual regulation. Collectively, he and his colleagues drew attention to issues of housekeeping that had long been more or less ignored in the social sciences and the humanities – energy transfer, adaptation, resilience, metabolism, and cybernetic regulation, to mention some of the buzzwords. This scholarship was a great leap forward, paving the road for important ethnographies focusing on environmental relations and concerns, providing a "thick" description of subsistence. Rappaport's work, however, had significant limits. While he was familiar with housekeeping tasks at several scales, having run a chain of hotels and restaurants before turning to anthropology, his approach was fairly narrow, focusing on a few households. Given this baggage, what could be the role of environmental anthropology and related fields in the current age, with its planetary connections and escalating, anthropogenic environmental change?

For many people, the massive scale of environmental change during the Anthropocene, the complex and poorly understood dynamics involved, and the distributive agency of the human and the more-than-human no doubt evokes a familiar idea of human paralysis when confronting alien Earth – a "cosmic terror" in Bakhtinian terms, "the fear of the immeasurable, the infinitely powerful . . . the terror that pervades ancient mythologies" (Bakhtin, cited in Last 2013: 66–67). Is it the fate of humans, and the study of humans, to lose

[1] Henry Fairfield Osborn Jr., *Our Plundered Planet*, 1948, Little Brown.

perspective when faced with the unprecedented heat and terror of the Anthropocene? How close can we get without getting paralyzed? Paraphrasing Hannah Arendt, I suggest that a broader notion of the Anthropocene captures what might be called the "new human condition," posing unprecedented challenges for governance and housekeeping. The current collapsing of body, society, and environment, in both biosocial theory and the "real" world, necessarily invites applying the governmental gaze throughout, from the cellular to the global level, embedding humans and other beings in ever-larger contexts across the different scales of the Anthropocene.

12.1 Concerns with housekeeping: "from the doors inward"

The English term "housekeeping," it seems, was developed in the sixteenth century. Usually it has denoted activities for maintaining a domestic household, but recently it has also been applied to computing (maintaining systems) and biology (housekeeping genes). Many historical and ethnographic accounts testify to more or less identical concepts and concerns related to the vitality of the house economy. For ancient Norse, for instance, humans and the land in which they were embedded represented an integrated universe of its own, an *óðal* (hence the German *edel*) requiring collective attention, maintenance, and care (Gurevich 1992: 178).

Another example is that of the "house model" or economy of livelihood of rural Colombia and Panama (Gudeman and Rivera 1990, Gudeman 2012). For Colombian peasants, the notion of the house (*la casa*) refers to a rural economic group and the manner in which it cares for its base or foundation (its wealth or capital in a broad sense). Caring for the base involves two projects, maintaining the "force" or "strength" (*fuerza*) of the base and "augmenting" or "increasing" it. Through both projects, rural Colombians seek to hold (*guardar*) household products "inside the doors" – in order to ensure sustainability, to use modern environmental jargon; "they practice 'householding' and 'housekeeping'" (Gudeman and Rivera 1990: 40). The house metaphor, here as in many other contexts, is pervasive and compelling:

> Within a culture ... some figures seem to hold a persistent central place, and the "polity-as-house" example suggests something of the persuasive power the "economy-as-house" metaphor still has in rural Colombia. (Gudeman and Rivera 1990: 41)

In the South American model, the house economy is a sustainable closed unit with a vital force of its own. In local terms, when the house is self-provisioning it acts "from the doors inward"; if, on the other hand, the base is not replaced through work it becomes depleted and the house falls into "ruins" (Gudeman 2012: 63–64). For Gudeman (2012: 64), this is "debasement," "a collapse more total than is suggested by our word 'bankruptcy,' for a house in ruins has no outside social supports or communal welfare." The threat of debasement is one of the characteristics of the global household; there is no outside support to draw upon to make ends meet. South American peasants speak of unique celebratory moments, perhaps once in a lifetime, as "throwing the house out of the window" (Gudeman 2012: 71). The Anthropocene doesn't allow for such luxury. An outside solution, from the doors *outward*, to the problems at hand would probably require a good deal of the Milky Way.

A similar house model, the ancient Greek notion of *Oikos*, the household of life, provided the root for the modern notion of "economies" and "ecosystems" and the names of their respective disciplines. In *The Politics* Aristotle argued that the products of agriculture, fishing, and hunting were "given by nature to all living beings"; "it is the business of nature to furnish subsistence for each being brought into the world ... Plants exist to give substance to animals, and animals to give it to men" (cited in Gudeman and Rivera 1990: 145). In modern terms, this would be "ecological economics," a house model of the house at whatever scale.

The definition, boundaries, and modern implications of house economies have, however, been the subject of some debate. In fact, an "*Oikos* controversy," launched by the German economist Karl Johan Rodbertus in the 1860s, has its respectable place in the house of economic history. The center of the debate was the question of whether or not the economies of antiquity – Greece in particular – were highly developed models of the modern market economy. Was the "stage" of the *Oikos* economy characteristic for antiquity overall or only certain primitive social formations, and what, if anything, might be learned from it? Weber, one of the influential contributors to the debate, doubted that *Oikos* had any relevance for modern society, arguing that it narrowly applied to the household of a prince or manorial lord.

Polanyi revolutionized the tired controversy, revitalizing the primitivist case in looking for "a mechanism for recreating in the complex societies of modernity the sort of overview that had existed in the household (*oikos*) economies of the past" (Dale 2010: 28). "The household," Polanyi argued, "is the smallest, the polis is the largest unit of consumption: in either case that which is 'necessary' is set by the standards of the community" (Polanyi 1967: 78). Rehabilitating Aristotle's sociological approach to the "embedded" economy and human affairs, Polanyi emphasized, much like modern South American peasants, community and autarchy:

> The group as a going concern forms a community (*koinonia*) the members of which are linked by the bond of good will (*philia*). Whether *oikos* or *polis*, or else, there is a kind of *philia*, specific to that *koinonia*, apart from which the group could not remain. *Philia* expresses itself in a behaviour of reciprocity ..., readiness to take on burdens in turn and share mutually. Anything that is needed to continue and maintain a

community, including self-sufficiency (*autarkeia*) is "natural" and intrinsically right. Autarchy may be said to be the capacity to subsist without dependence on resources from outside. (Polanyi 1967: 79)

Polanyi could not have imagined the relevance of these words half a century later for the global discourse of the Anthropocene. Again, the house model seems pertinent, with its focus on good will and material limits. Perhaps it is time for another *Oikos* controversy, on a radically different spatial and political scale.

Recent debates on the *Oikos* have centered on production, gender, and agency. While economic theory originally defined economic space in terms of the broad spectrum of activities in the domestic household, it later tended to restrict its definition of "productive labor," a central economic term, to activities taking place *outside* the domestic sphere. Arguably, the notion of the household is flawed on two scores. For one thing, the house model was the invention of agrarian discourse. Given the mobility of humans – both prior to agriculture when families and communities regularly moved from one camp to another and nowadays in times of excessive travel and migration across the planet – the stable house erected on a foundation may be somewhat misleading. Also, and more importantly, the house is rarely, if ever, completely egalitarian. While there may be reasonable equity in the Colombian household (Gudeman 2012: 65), ancient Greece was the opposite with its radical separation of free persons and slaves and of men and women. Inequality in the global household, particularly between North and South, is a major barrier to joint pragmatic solutions to environmental problems. One of the contradictions of the Anthropocene is the fact that despite the frequent reference to the unqualified "us," the human collective charged with correcting its devastating footprint on the planet, much of humanity is denied citizenship and subjecthood, the homeless migrants and refugees of the age of globalization; in the terms of postcolonial theory, this is the new subaltern (Chakrabarty 2012: 4).

Despite the flaws of the *Oikos* concept, the family hearth has to be sustained – irrespective of the frequency of movement, in the course of struggles for justice, democracy, and equity – to avoid total "debasement" in the South American sense, to secure decent living conditions and the necessary flow of vital energy for the generations to come. It seems safe to assume a family resemblance of processual house*keeping* concepts – references to the act of maintenance – more or less across languages and throughout human history (Carsten and Hugh-Jones 1995). In Marxian theory (Godelier 1986: 6–7), the conceptual pair of *Überbau* and *Grundlage* seems to capture the essence and unity of the Colombian house, the Scandinavian *óðal*, and the Greek *Oikos*, of both their base and the life form they represent and sustain.

One of the pressing tasks on the environmental agenda is to identify the nature and novelty of the current condition for humanity as a species.

One source of insight is Hannah Arendt's *The Human Condition* (1958), a treatise of political philosophy on the various forms of human activity that relate people to one another and to the material world. Arendt suggested that humans' being-in-the-world had been disrupted in the modern age, resulting in alienation from the common, artefactual human world and from nature. Arendt, it is true, made a critical distinction, in a classical fashion, between the space of politics, the *Polis*, and the household, the *Oikos*: "It was only upon entering the *polis* and leaving behind the cares of the *oikos*, upon freeing oneself from the cares of sheer life, that the citizen could pursue good life" (Collier and Lakoff 2005: 26). Nevertheless, she argued that times had changed: "In the modern world, the two realms indeed constantly flow into each other like waves in the never-resting stream of the life process itself" (Arendt 1958: 33). In a sense, the "rise" of the household into the domain of politics – for Arendt "an essentially modern phenomenon" – has now been completed, with the "biopolitics" of both the Anthropocene and life itself.

In his discussion of the relevance of Arendt's politics of place for discussions of the household of life, Macauley suggests that Arendt rarely focused directly on the environment and at times she was too committed to the ancient Greek tradition to adequately deal with the ethnographic realities of the contemporary; nevertheless, she was deeply concerned with dwelling and rootlessness, issues that have direct relevance for current discussions of accelerating human impact on the earth:

> In a day when much of humanity is more concerned with exploring "outer space" than recovering our own sense of earthly place, this kind of thinking can help us to reconceive our relationship with nature and the world ... In the end, it may be said that Arendt did not so much initiate a dialogue with the earth itself as she suggests ... but rather kindled a valuable inquiry into *the origins and meaning of our changing relation to and transformation of our given home*. (Macauley 1996: 126; emphasis added)

Modern biological and economic theory has tended to reduce the rich and integrated world of the ancient household of life to the "natural" domain, stripping it of the "social." As Schabas (2005: 12) points out in her analysis of the conceptual roots of economic thought, social as well as natural scientists "have implicitly agreed to divide the world into two parts. When physicists today think of the world they investigate, it is one with all the social institutions stripped away ... Economists, in parallel fashion, have come to adopt a domain of discourse that is similarly segregated." Such fragmenting, which has characterized anthropology with its segregation of the social and natural aspects of *Homo sapiens*, no longer sounds theoretically defensible; the application of concepts such as nature/culture (Haraway 2008) and biosociality (Ingold and Palsson 2013) increasingly seems unavoidable for rendering the refashioning of the human being, its house, and life itself.

Before Adam Smith placed markets at the center of economics, James Steuart (1713–1780), one of the pioneers of modern economics, used the image of the house as a metaphor for the national political economy, and long before him the mercantilists drew upon a similar model (Gudeman 2012). Given the collapsing of the natural and social in the modern age and the immediate concerns of environmental change, it seems pertinent to draw upon the ancient notion of the *Oikos* and to vastly expand the meaning of housekeeping, to humanize the discourse of the "earth system" and to avoid the nature/culture dualism inherent in the notion of planetary boundaries (Rockström *et al.*, 2009). As Nelson (1993: 33) remarks, keeping in mind the notion of the house in both the words "economics" and "ecology": "Economics could be about how we live in our house, the earth."

12.2 Anthropogenic change

Concerns with large-scale housekeeping necessarily escalate with growing evidence of human impact, as our footprint on Earth begins to match those of geological forces. It is patently clear by now that the "natural" climate of the globe has a lot to do with "artificial" tangible and intangible human services and goods. Some Arctic volcanic eruptions and earthquakes are even attributed to human activities; a recent study on Iceland's Vatnajökull ice cap suggests that melting glaciers can increase volcanic activity over timeframes that are relevant to humans (Pagli and Sigmundsson 2008). As a result, it is being argued (Steffen *et al.* 2011), planetary history has entered a new, emerging epoch, the Anthropocene – a successor to the last interglacial epoch, the Holocene. Similar concepts range from anthropozoic era, the Homogenocene, and the catastrophozoic era to the Great Acceleration. While the insight that humans have become one of the dominant factors in shaping the globe is not new, the Anthropocene concept is one of the most influential concepts attempting to capture this insight.

The discourse on the Anthropocene, highlighting as it does growing evidence of the exceptional role of humans in the refashioning of life on Earth, is not free from critique and debate. One issue is that of definition and identification. The idea of allocating the Anthropocene a respectable place in the global standard of geological periods has not remained uncontested. Geologists have actively debated the exact onset of Anthropocenic change, the stratigraphic legalities of the Anthropocene, whether the term would meet their strict protocols, how to detect the relevant signatures in the geological record, and so on (Showstack 2013). While other geological epochs have been defined in terms of the "golden spikes" of rock formation, geologically speaking the "rocks" of the Anthropocene do not exist yet. As a result, the relevant spikes for marking Anthropocenic time should not, and cannot, be looked for in fossil sequences (Kolbert 2013).

It is now expected, nevertheless, that the so-called International Commission on Stratigraphy will accept the Anthropocene as a new geological age at a meeting in 2016. While the debates and arguments of the Commission are not trivial, the narrow confinement of anthropocenic discourse to the community of "science" and the notion of the geological record fails to illuminate the various signatures of human concerns and activities on the global scale. Geological time itself, in the conventional sense, seems out of time, fatefully informed by the ancient European semiotics of the book of nature. Geological fundamentalism – the equivalent of the Periodic Table of the Elements in chemistry – doesn't seem to resonate with the fleeting, humanized Anthropocene. What is missing is genuine integration of social and environmental theory appropriate for the times.

Anthropologists, philosophers, and social historians are unlikely to bother with legalizing and standardizing their lists of ages and periods. For those familiar with "chronic" debates on the relativity of timing and chronotypes (see, for instance, Bender and Wellbery 1991) and the definition and demarcation of periods – Antiquity, the Long Eighteenth Century, the Middle Ages, the Renaissance, and so on – concerns with the making of epochs need not be surprising. The term Middle Ages was only invented in 1469 to emphasize the inferiority of the Dark Age, the period prior to the Renaissance; it is high time, Le Goff (1988: 10) argued, "that we let the air out of the inflated concept of the Renaissance." While the "Anthropocene" has only been around for roughly a decade, it is probably already more inflated than any other term in the entire history of the timing of epochs and chronotypes.

What matters, however, is the pragmatic usefulness of marking time, its implications for our understanding of the human in the current age. There may be good reason, I think, to speak of *four* Anthropocenes, not just one. The longest one, the *very long* Anthropocene, would be the one marked by the early use of fire. It seems possible that in prehistoric times more than one species of primates had significant environmental impact. Neanderthals, we now know, extensively used fire (Daniau, d'Errico, and Sánchez Goñi 2010) and their role has probably been underestimated. Given such a time-scale, at the risk of trivializing the issue, perhaps one should speak of a "Primatocene" rather than an Anthropocene. The *long* Anthropocene, in contrast, would be the era signified by the beginning of agriculture, with its large-scale transformation of landscapes and political systems. So far, most of the writings related to the Anthropocene, however, assume a *short* era, suggesting that it started in the late eighteenth century in the wake of colonial expansion and capitalism, with rapidly growing combustion of fossil fuels (Tickell 2011) and "ecological imperialism" (Crosby 1986). Finally, the *immediate* Anthropocene is that characterized by human awareness of anthropogenic change. Indeed, the era of the Anthropocene does not just imply conflation of the natural and the social, also it suggests radical change in perspective and action in terms of human awareness and responsibility.

The Anthropocene, it may be argued, could be seen as an anthropocentric construct. There are good grounds, after all, for speaking of *distributive* agency to emphasize that the Anthropocene is not the result of *Homo sapiens* (or primates) acting in isolation; it is only made possible through a diverse network of organic, technological, cultural, and geological entities (Connolly 2011). As Latour (2013) points out, the somewhat surprising *comeback* of *Anthropos* in the postmodern, post-human era represents a major conceptual feat, returning humans to the driver's seat from which they were recently expelled at a time when social theorists are busily granting agency to everything living. However, the most striking feature of the Anthropocene is that apparently it is the first geological epoch in which a defining geological force is *conscious of its geological role* (Palsson *et al.* 2013). "So maybe," Szerszynski (2012: 171) argues, "the Anthropocene in all its geohistorical specificity really starts when humans become aware of their role in shaping climate, and this awareness shapes their active relationship with the environment." This awareness has profound implications for politics and human responsibility. The immediate Anthropocene which began by the middle of the last century is obviously of greatest relevance for any discussion of environmental concerns and planetary housekeeping. In the language of Osborn in 1948, "it is man's earth now." Leaving aside the gendering in Osborn's language, how should the figure of "man" and human agency be understood?

The circulation and use of the term Anthropocene has escalated exponentially, much like many of the curves describing the rate of anthropogenic change, including global warming and melting glaciers. Googling it in late 2013 yielded one million results. Should we let the air out of it, much like Le Goff did for the "Renaissance"? Not necessarily, in my view, and, in any case, who are "we"? Drawing upon Thoreau's critique of the egocentric naming of "Flint's Pond," Crist (2013: 142) suggests the label of the "Anthropocene" reflects a similar "poverty of nomenclature" and that accepting it would be "to bow once more before the tedious showcasing of Man," a managerial mindset and a Promethean self-portrait. While the geology of the Anthropocene is only prospective, a reasonable hunch about where things are going, the ontological problems and the tensions of the universal and the particular remain, I suggest the anthropocentric notion of the Anthropocene may turn out to be a powerful *pre-emptive speech-act*, fueling responsible attempts to avoid disaster, what *might* and *is likely to* become. Admitting the flaws and biases of the label of the Anthropocene, Clark emphasizes its positive aspects; while current environmental problems represent a mega-disaster, undermining our opportunities for making sense of what is going on, "the Anthropocene is as much about the *decentering* of humankind as it is about our rising geological significance" (2014: 25; emphasis in the original).

12.3 Icarus in the heat: the plowman and the splash

In the Anthropocene, humans are the driving force of major environmental accelerations that seriously threaten life as we know it. At the same time, what life itself is understood to mean has been increasingly destabilized in the wake of massive intellectual and practical changes involving a complex array of theoretical, technical, and empirical innovations, including those of epigenetics, systems biology, and microbiomes. The task of housekeeping, as a result, involves both caring for our bodies, our families and communities, and everything that surrounds us. No wonder that life itself has become the center of debates on governance and the relative merits of public trust, markets, and states. Despite these developments, the routine critique of the ontology of the dualism of nature and society (Descola and Palsson 1996, Hastrup 2013), and increasing emphasis in both social and biological theory on mutualism and a variety of relational concerns (Shoreman-Ouimet and Kopina 2011), the interdisciplinary literature on governance often presents "resources" and people as clearly separate domains, with the latter defining and appropriating the former, through ever-more innovative avenues.

Human awareness and responsibility, as already mentioned, often engenders terror and paralysis. Perhaps the cosmic terror of the Anthropocene is reminiscent of the predicament of Icarus in Greek mythology. Presented with wings of feathers given to him by his father, Daedalus, a talented craftsman, Icarus decides to explore the sky, normally the privileged domain of birds and gods. Despite his father's warnings, he eventually flies towards the glowing sun. As he gets close, the wax that has secured the feathers begins to melt, and Icarus falls helplessly into the ocean. This is a theme explored in Bruegel's representation of the myth of Icarus in his famous painting *Landscape with the Fall of Icarus*. Here, Icarus's legs can be seen in the water, with the sun half-set on the horizon and peasants calmly continuing to work in the field despite the fall. A poem by W.H. Auden, "Musée des Beaux-Arts," nicely describes the spectacular fall and the indifference of those who were bound to witness it:

> In Bruegel's Icarus . . .: how everything turns away
> Quite leisurely from the disaster; the ploughman may
> Have heard the splash, the forsaken cry,
> But for him it was not an important failure; the sun shone
> As it had to on the white legs disappearing into the green
> Water; and the expensive delicate ship that must have seen
> Something amazing, a boy falling out of the sky,
> Had somewhere to get to and sailed calmly on.
>
> (Auden 1940)

Perhaps the melting of the wings and the splashing of Icarus's body can be seen as a reminder of the entanglement of organisms, their epigenetic regulation, and their dependence on the house and its vital circulation of energy. Also, it draws attention to our ability, or inability, to act quickly and responsibly to the terror of the Anthropocene. Will the wax on our wings melt as our flight continues, sending us into the sea with a splash without any historical record? Alternatively, we might ask, in a more optimistic mood: How can we theorize our flight, managing and redirecting our course? For that historic task, a thoroughly transdisciplinary effort is essential. It seems vitally important to come to terms with both the terror and the heat involved and the scale of the human footprint.

Perhaps the most pressing task on the theoretical and political agenda involves navigating the transition to a full-blown "Anthropocene society." The Anthropocene is not just a new geological epoch; it redefines the very nature of the geological by marking it as a domain that includes intentionality and meaning. It seems time to address the issues involved with consistent, transdisciplinary effort, assuming human responsibility, social justice, and public trust. Significantly, the inclusive term "environmental humanities" has been gaining ground in several contexts. The great challenge now for anthropology and related disciplines is to ethnographically document the unfolding of the Anthropocene in the broad sense and to theorize its significance. This involves exploring the extent to which the human condition, as analyzed by Arendt, has changed in the Anthropocene era and the nature of this change (Chakrabarty 2012; Palsson *et al.* 2013; Clark 2014). Virtually all social sciences and humanities disciplines have at some point taken an interest in human–environmental interactions. Indeed, they are critical for thinking the Anthropocene and the environment. While the notion of the "environment," a key term in Anthropocenic discourse, is often taken for granted, it has a history that tends to elude us. The two terms of environmental "expertise" and "environment," Sörlin (2013: 15) points out, have been "co-produced as a kind of double helix so that the modern usage of the 'environment' is essentially constructed by those that claimed expertise on it and also provided the advice." Gibson and Graham suggest that what is needed is a new ethics of care, relating to the global world as one does to a family; can we extend our solidarity, they ask, to the more-than-human, to other life forms and life in general? "If we can," they conclude, "that would certainly usher in a new mode of humanity and a new form of belonging" (Gibson-Graham 2011). This will mean reorganizing our own house in a radical sense – a central element of the housekeeping of the Anthropocene. Because what currently counts as environmental is also social (or, in some accounts, biosocial or natural/cultural), humanity's knowledge enterprise needs to return its attention to social theory and the humanities for reframing *Anthropos* for the modern context.

13 Afterword

During the past half a century or so, both life and anthropology (and related themes and disciplines) have taken sharp turns. In her perceptive review of the state of anthropological theory, "Anthropology since the Sixties," Ortner (1984) nicely captured the decline of the three schools dominant in the 1950s (structural functionalism, cultural and psychological anthropology, evolutionary anthropology), the new frameworks that emerged during the 1960s (symbolic anthropology, cultural ecology, structuralism), how each of them drew upon and differed from their predecessors, and how they collectively, in turn, generated new theoretical avenues during the 1970s (in particular, several kinds of Marxism). Reflecting on the landscape ahead, she anticipated a new emergent "key symbol of theoretical orientation" (Ortner 1984: 127) – namely, practice or praxis, partly through the influence of Bourdieu (1978). Ortner was right, but only up to a point.

13.1 Styles and context

Any convincing account of the development of anthropological theory since the eighties is bound to give substantial weight to the shift from structure to agency and to the various ways in which practice theory has been developed and used during the subsequent decades. Ortner's claim, on the other hand, about the fuzziness of community boundaries, the quiet and peaceful coexistence of "small coteries," and the absence of "stirring arguments" in the early 1980s failed to foresee the rough and divided road ahead:

> The field appears to be a thing of shreds and patches, of individuals and small coteries pursuing disjunctive investigations and talking mainly to themselves. We do not even hear stirring arguments any more ... We no longer call each other names. We are no longer sure of how the sides are to be drawn up, and of where we would place ourselves if we could identify the sides. (Ortner 1984: 126)

The same year that Ortner's article was published, a group of scholars gathered in Santa Fe for a ground-breaking theoretical discussion that later materialized in *Writing Culture* (Clifford and Marcus 1986). Signaling the birth of a long-standing and heated debate on the postmodern that would split the community into rather clearly demarcated camps, *Writing Culture* illustrated the difficulties of anticipating theoretical currents that set the stage for years to come.

During the 1930s, years before the concept of scientific paradigms gained currency, Fleck (1979: 142) pointed to the idea of changing "thought styles." If thought styles name the world, generating perceptions that "stick," there has to be a collaborative and receptive audience; namers and context, so to speak, coproduce those styles. What, then, have been the main forces channeling anthropological interests, shifts in perception, and persistent theoretical scrutiny? Several radical changes in intellectual culture and geopolitics have helped to shape anthropological theory. One of the key changes involved was that of the postmodern turn that would inform the discussion and framing of many key issues, including ethnographic authority, the constitution of ethnographic facts, the proper manner of writing ethnographies, and, last but not least, the decline of grand narrative.

The prime example of grand narrative in anthropology and the humanities is structuralism, which also inspired theorizing in several other fields, including linguistics (Roman Jakobson, Noam Chomsky), the psychology of cognitive development (Jean Piaget), and biology (François Jacob). Aided by mathematical language and precision, the discovery of previously hidden structures and units (phonemes, mythemes, genes) was assumed to explain just about anything. Anthropology, then, had a powerful theory that many other disciplines could emulate and adapt for their own purposes. In recent decades, however, partly due to postmodern critique of totalizing theories, structuralism has increasingly been under attack. Structuralist perspectives, however, remain strong in some fields, in particular the gene talk of molecular biology and biomedicine, but even here the semiotic systems presupposed have increasingly been subsumed by the biosocial turn – invaded by relations and context, by challenges posed by, for instance, epigenetics and developmental systems theory (Krimsky and Gruber 2013).

To some extent, postmodernism underlined an anthropological Mid-Atlantic Rift by the turn of the century, with its tectonic friction and divisions. It was not, in fact, a uniform thought style with the same perceptions and effects in different academic contexts on either side of that rift. Godelier (2009: 2, 19) warns that the collective work of Lyotard, Derrida, Foucault, and Deleuze, known in the United States as "French theory," was "a purely American invention, since France recognizes no such unified body of thought," adding that one should not put "Americans on trial," and "that . . . conflating all those who have rallied to or are accused of having cheered 'Postmodernism'" would not make sense. There were, indeed, important North American strands, including that of the interpretive anthropology of Geertz.

Postmodernism, no doubt, had a useful impact in that it sensitized the anthropological community and related fields to the role of the ethnographic scribe, highlighting the need to situate accounts and, more importantly, to continue to experiment with methods and representations. Significantly, the modernist notion of ecological anthropology that was popular in the 1970s and the 1980s, which tended to separate human discourse from the systems observed, seems to have been replaced by the more open-ended label of "environmental anthropology," emphasizing situated accounts, the perspective of dwelling, and the unity of humans and that which surrounds them (Ingold 2000).

In the long run, on the other hand, the appeal of radical postmodernism began to fade. After all, it was argued, some kind of grand narrative might be unavoidable, even essential, for responsible scholarship, given the pressing global problems faced by humans and life itself. Would environmental anthropology, for instance, be of any use if it only deconstructed discourses on climate change, and would it be socially and environmentally responsible to abandon any attempt to establish the conditions of the biosphere and to inquire into the implications for anthropological subjects? And has postmodernism, which by now has been around for almost a century, not had a grand narrative of its own, an ironic and cheerful nihilism?

At more or less the same time that postmodernism was gaining force, globalization and neoliberal politics settled in, paving the way for the new economy of Late Capitalism, an economy characterized by, among other things, venture capitalists, fluctuating markets, Internet technology, and digital trading (Fisher and Downey 2006). These events generated theoretical interest in a series of new and renewed themes, such as cultural flows, spatial formations, property rights, and the nature of the political economy. For some analysts, virtualism represented a new political economy that conflated the domains of economy, society, and the community of economists modeling the market: "Perceiving a virtual reality becomes virtualism when people take this virtual reality to be not just a parsimonious description of what is really happening, but prescriptive of what the world ought to be: when, that is, they seek to make the world conform to their virtual vision" (Carrier 1998: 2). Not only did these developments echo both the postmodern condition and K. Polanyi's early work on "fictitious commodities," they also seem pertinent for theoretical understanding of more recent dramatic changes in both the world economy – global financial crises, marketing bubbles, and collapses of banks and national economies – and the political economy of life itself.

For several scholars, a return to Marxian theory – suppressed and marginalized during the heydays of postmodernism, neoliberalism, and the Cold War – provides an important avenue for understanding power, inequalities, biopolitics, and the human condition more broadly in Late Capitalism.

Patterson (2009: 159–160) suggests, for instance, that Marx would be relevant for a number of modern themes, including the historicity of human beings, capitalism, and transformations on an increasingly global scale. Keeping in mind Polanyi's early work on the fictitious and Carrier's notion of the virtual, it is rather saddening to see that anthropologists, with notable exceptions (for example, Tett 2009), failed to detect and disentangle the recent financial bubbles until the collapses they implied had eventually occurred. In the wake of the recent crises, it seems, virtualism needs to be dissected and exposed and placed firmly on the agenda, highlighting the importance of theorizing the transformations of capitalist economies and their implications for culture and society, locally and globally, and all kinds of anthropologies of life.

13.2 Disciplining thoughts

The classification of academic fields and disciplines is often seen as somehow natural or predetermined, as a final state of affairs bound to be revealed eventually through the processes of history. In C.P. Snow's classic formulation, the gap between the "two cultures" – between the natural sciences, on the one hand, and, on the other, the arts, humanities, and social sciences – was self-evident (Snow 2001). Increasingly during the past few years, however, academics have seen this binary construction as masking a much more complex field. Academic fields and disciplines are relatively recent and unstable phenomena established and modified in particular historical contexts. The conditions for academic practice continue to change, and, as a result, so do disciplines and their division of labor. At the same time, new fields emerge in response to new developments. A blurring of genres seems inevitable; the future academe may be multicultural and transdisciplinary. During the twentieth century, spectacular developments in genetics and related fields, including the so-called new genetics, not only elevated molecular biology and the life sciences to the status of Big Science but made biology arguably more important than physics. Where does this leave anthropology, a field traditionally divided along the biological–social axis? Will the new genetics and associated theoretical and technological developments signify a reshuffling of the anthropologies taught and practiced in the twenty-first century? Will they lead to an even more hybrid and lively assembly of "anthropological" paradigms and practices?

Recent research in the fields of anthropology, cultural theory, sociology, as well as history and philosophy of science, has readjusted the notion of knowledge as something that is already there and only needs to be retrieved, towards the notion of knowledge as something that is produced through social practice. Not only has the modernist project, with its notions of truth and grand narrative, increasingly been challenged, the academic landscape of fields and disciplines in itself is tense and unstable. The changes now taking place may be just as spectacular as the changes represented by the Renaissance and Enlightenment. If disciplines, subdisciplines, and fields of inquiry arise, develop, and disappear, much like natural species in evolutionary theory, how should one represent this process? How is it disciplined (in the dual sense of controlling and fragmentation), what validates the candidacy of a discipline or a field, and what would be the languages and metaphors most appropriate for the theoretical understanding of current and future developments? In the past, spatial metaphors were commonly used to talk about disciplinary fragmentation: borders, zones, areas, fields, provinces, pastures, and territories. Recently, new hybrid zones of intellectual discourse *bordering on* different fields of science have appeared at a growing rate. Indeed, transdisciplinarity has its advantages, as much creative work develops at the interfaces of established fields of scholarship. At the same time, some transdisciplinary issues and perspectives that used to represent a specific voice or platform in several fields, notably race and gender, often seem to have been co-opted, mixing with and reshaping the fields in question. This is evident in anthropology, where many feminist perspectives formerly at the critical margin have moved to the core.

Clearly, Snow's idea of the two cultures seems increasingly archaic, given the current conflation of all kinds of domains and the destabilization of the dualisms of the past. Keeping in mind the conflation of the domains of the natural and the social and the radical culturing of life with the new genetics, the demarcation of sociocultural anthropology, on the one hand, and, on the other, biological and physical anthropology seems misguided (*contra* Segal and Yanagisako 2005). It seems likely that anthropology will be redefined, in terms of both its internal divisions and its relations to other disciplines. While there is often strong resistance in academies and universities to erasing disciplinary boundaries, quite possibly the current "blurring of genres" will make academe *post*-disciplinary.

Not only have there been several roads out of the writing culture of Santa Fe, also in recent decades anthropological theory has been informed by several other nodes in the emergent network of events and ideas. Some turned out to be dead ends, others have recycled earlier themes, much like the theoretical schools charted by Ortner (1984), while still others have opened up new ethnographic and theoretical vistas for the future. Anthropologists will continue to debate, hopefully without calling each other names, what counts as dead-end and what represents a new and promising vista. Given, however, both the refashioned human condition in the contemporary world and the contested understandings of some of the core orientations and key concepts of anthropology, anthropology and related disciplines need to do extensive homework on theory, concepts, methods, and disciplinary formation and collaboration. Significantly, as we have seen, a biosocial turn is characterizing academic agendas to an unprecedented degree, across both the social and the life sciences.

Method, of course, is informed by social context and social theory. Whatever its advantages, the classic, single-site model of ethnography demands important qualifications. At least two major methodological trends have taken place in recent decades. The first is that of multi-sited fieldwork emphasizing cultural and material flows, spatial connections, and different scales of observation, moving near and far, from the center to the margin and back. Multi-sited fieldwork is at least partly a byproduct of globalization and changing conditions of fieldwork in the postcolonial era. Secondly, collaborative methods have been gaining ground. Ethnographers often speak of collaboration in relation to the entire research process, from research design to publication.

This involves expanding the community of modelers, blurring the boundary between the observer and the observed in the coproduction of knowledge. One example is Gudeman and Rivera's (1995: 245) emphasis on the making of ethnography "*with* a people": "Doing ethnography is joint work, teamwork, not a discovery, and not an interpretation of a preexistent, fixed culture." For Lassiter (2005: 72), the radical move to collaboration has much to do with the shifting context of fieldwork and the growth of multi-sited ethnography: "In sum, the 'new ethnography' potentially moves collaboration from the taken-for-granted background of ethnography to its foreground." No doubt, collaborative work is also informed by the growing impact of practice theory and postcolonial studies, granting agency to the people formerly identified as passive informants. An associated and still relatively underdeveloped methodological development is that of studying up, rather than down or sideways, in science studies of laboratories and high-tech teams. Interestingly, in such contexts the tone of humility characteristic of collaborative ethnography is often submerged, if not entirely lost.

Neither multi-sited ethnograpy nor collaboration is entirely new. While many of the early ethnographers tended to stick to a single site throughout a field period, seeking to develop a broad perspective, some of them repeatedly moved from one site to another. Likewise, collaborative approaches draw upon earlier experimenting with dialogue and consultancy in feminist scholarship and critical ethnography. No doubt, experimenting with multi-sited approaches and collaborative forms of publication and dialogue will continue, but these trends are no longer seen as experimental; rather, they represent emergent norms that young scholars tend to see as appropriate to the conditions of ethnographic work in the contemporary world.

13.3 The house and the body

The discipline of anthropology is part of the ongoing negotiation of the boundaries between the social and the biological, given its conventional fragmenting of *Anthropos*. Keeping in mind that the notions of "biology" and "the body" have been radically socialized, at the same time as the notions of "social theory" and "society" have been thoroughly embodied and materialized, it seems pertinent to rethink *Anthropos* and the discipline of anthropology. Attempting to make sense of the political economy of modern biotechnology, including the fragmenting of body parts and the labor processes involved, a series of scholars (for example, Lock and Nguyen 2010) have revisited the writings of Marx and Foucault. Quite possibly, the growing awareness of the conflation of the natural and the social through studies of the post-human and contemporary biopolitics will be one of the key forces shaping anthropological theory, and much of academe, in the decades to come.

The house, the body, and the environment are intimately connected; the house serves like an extra skin mediating between persons and the context within which they are embedded. Ethnographic evidence shows that houses are often thought of as bodies (Carsten and Hugh-Jones 1995: 2). Housekeeping, then, is frequently a matter of both constituting bodies and environments. In the South American model, the flow of vital energy connects the house and human bodies. Just as the energy of life – *vis vita* – drives the household, it sustains the body, in fact all living things. As one peasant put it, "[i]n working you are using the strength of food ... When you eat, you replace this strength ... This goes on and on in the body. People suffer when they do not have enough force to work. They are using up their health" (Gudeman 2012: 60). Given a housekeeping model of the Anthropocene, this "vital" connection between person and planet, organism and environment, needs to be restored.

Charles Darwin may provide a lead. In a well-known passage in *The Origin of Species*, he presented the notion of entanglement somewhat at odds with his main thesis:

> It is interesting to contemplate an entangled bank, clothed with many plants of many kinds, with birds singing on the bushes, with various insects flitting about, and with worms crawling through the damp earth ... There is grandeur in this view of life, with its several powers. (Darwin 2008: 210–211)

Human entanglement is characterized by the complex biosocial relations within which we are embedded. Food, of course, provides one of the key avenues. As Probyn (2013: 289) suggests, "[a]cross a number of scales, eating ties us into global-local geo-political webs: the ecological landscape of food-human and non-humans."

Nutritional environments (the kind and quantity of food) – early environments, in particular – condition the possibility for gene expression, potentially affecting our health and well-being. Such "epigenetic" regulation seems to be prevalent in the human genome. Niewöhner (2011: 281) suggests that social and cultural anthropologists may benefit from the challenge of what he calls "environmental epigenetics," to "rethink the role of materiality for the patterning of social practice." According to the field of nutritional epigenetics, food transforms the organism's being, leaving an imprint on the body (Landecker and Panofsky 2013). A similar idea seems captured in the South American house model; physical (phenotypic) resemblance despite different (genetic) ancestry is accounted for by "sharing the strength of the earth" (Gudeman 2012: 65). In light of the profound changes that have taken place in the understanding of bodies and their "environments," to what extent do they present new problems of governance during the Anthropocene?

Along with some other body issues, genomic and microbiotic material seems to invite new property dimensions and considerations – new kinds of housekeeping. With the new genetics, environmental epigenetics, the development of biomedicine, and the expanding production of biocapital, the very notion of the "biological world" has been destabilized as "nature" is increasingly subject to artificial, human, and social refashioning (Rose 2005). Moreover, the possibility of zooming in on the micro-world of cellular material inevitably destabilizes common notions of the biological. This in turn demands a rethinking of the governance of biological commons and raises important questions about the relevance and applicability of new institutional frameworks.

13.4 Environmental politics and theory

Discussions on how to address what now counts as "environmental" concerns have a long history, especially in political theory. William Lloyd wrote his tract "On the Checks to Population" in 1833, exploring the problem of overstocked pastures in the commons of old England with a conceptual framework that later inspired much work on commons (Lloyd 1977). Until the 1980s, however, the commons had not been addressed in the context of consistent comparative and empirical work juxtaposing speculation and theory, on the one hand, and actual regimes, on the other. At that point, several fields of scholarship – in particular, anthropology, ecology, economics, and political science (see, for instance, McCay and Acheson 1987) – collectively established a new interdisciplinary domain focusing on the cultures, practices, and institutions associated with the governance of commons.

For early political economists – Adam Smith, David Ricardo, Karl Marx, and some others – labor activities were, by definition, directed at the extra-somatic, external world. Thanks to their labor activities, however, humans are now able to reproduce their own bodies, as part of the "inorganic body" of nature, turning themselves into a "resource." Not only have the capacities of the body been fragmented and turned into instruments for production, redefining both human labor and human bodies, the sites of labor and production have increasingly been separated with the growth of the World Wide Web, network society, and virtual migration. "Paradoxically," Aneesh (2006: 2) notes, "the new space of transnational labor has reversed its relationship with the worker's body. Rather than move the body across enormous distances, new mechanisms allow it to stay put while moving vast quantities of data at the speed of light." This, too, invites commons issues and concerns. Governing the "house," as a result, has become vastly more complex than classical theory suggests.

Without doubt, the new interdisciplinary effort to address questions related to governance was partly informed by the growing environmental problems of late modernity, including those posed by rapidly expanding human populations, ever more efficient technologies of extraction and exploitation, and the near collapse of entire ecosystems and animal populations, especially fish. More than any other single recent work, Ostrom's book *Governing the Commons* (1990) carved the new interdisciplinary domain of the commons. In the process, our world of concern was expanded. "Think galactically, act locally!" is not just a Star Trek thing on the Internet, there are serious concerns with the commons of both the globe and celestial bodies (with rubbish in outer space and cultural heritage on the moon).

In recent years, political and environmental theory has been zooming *out*, emphasizing the interconnected world of Gaia and the detrimental "anthropocenic" signatures of humans on the global front. These concerns, however, are rarely connected and compared to those pertaining to the organismic and cellular level (Palsson and Prainsack 2011). A comparison of some recent works helps to illustrate the point. In her book *Reframing Rights: Bioconstitutionalism in the Genetic Age*, Jasanoff (2011: 3) usefully discusses the problems of entitlements in the context of life itself, pointing out that "periods of significant change in the life sciences and technologies should be seen as constitutional or, more precisely, *bio*constitutional in their consequences. Revolutions in our understanding of what life is burrow so deep into the foundations of our social and political structures that they necessitate, in effect, a rethinking of law at a constitutional level." Interestingly, however, despite her emphasis on the bioconstitutional, Jasanoff's book makes no reference to E. Ostrom or her followers. Nor do Ostrom's books on the commons pay any notable attention to the level of organisms, cells, and genes. Her book *Understanding Knowledge as a Commons: From Theory to Practice* (Hess and Ostrom 2006) only extends the commons perspective to the domain of information and the Internet. Her earlier book *Understanding Institutional Diversity*, which lists a number of cases and contexts for exploring the "building blocks" (Ostrom 2005: 9) of institutions, similarly makes no connection to the context of genomics and biomes. Given the growing recognition of human–environmental entanglements in the Anthropocene, including the development of epigenetics and microbiomics, it seems time to consistently link the levels of nano and giga.

The metaphor of the wind may be instructive to capture the problems of managing the flow of vital energy through bodies and biomes. As Evans and Reid (2014: chap. 4) suggest, "atmosphere" penetrates and profoundly influences our bodies. While the wind is best described as movement and force, not as chemical substance, it can be regarded as "fugitive" resource interacting with landscape to produce, among other things, bodily sensation and energy. Lifshitz-Goldberg (2010) suggests the wind is an example of semi-commons where private and common property overlap and interact. Not only do wind farms necessarily draw upon the features of the terrestrial landscapes of public plains and cities, also they produce a so-called "Shadow Effect"; once the wind has encountered a mill or a turbine its strength and energy content is reduced, and, as a result, other users in its path will be somehow affected. Similar concerns need to be raised and discussed for the context of the "winds" and "shadow effects" of epigenetics. Our porous bodies, then, and their internal and external networks and relations are becoming center-stage.

As long as one exclusively focused on the internal relations of cells and bodies, the issue of genomic governance seemed to resist the gaze and reasoning of political theory traditionally focused on the external biological world. Now, however, it seems we are back in a rather familiar terrain; microbiomes and epigenetic processes, after all, invite standard questions about distributive issues, rights to meaningful participation, social justice, and intergenerational fairness. One driving force behind the growing governance discourse of the late twentieth century was the advancement in many contexts of corporations, the "new economy," the free-market, virtualism, and neoliberal movements and governments. These were epidemic developments, no less straddling than migrating fishing stocks, ignoring conventional disciplinary boundaries and academic territories. Alternative governance models are needed for dealing with the global environmental crisis and the embedded body. In order to govern bodies and genomes it is necessary to both manage access to certain parts of our environmental commons (clean air and water, in particular) and to *avoid*, minimize, or adjust the impact of some others (toxic substances, radiation, and microbiomes, for example). These issues are already central concerns.

13.5 Key questions

Given the preceding discussion, the social sciences and the humanities need to play an important role in transdisciplinary research on the theoretical and political frameworks invited by the Anthropocene, posing new research questions and raising new issues that require new ways of thinking and acting. Among the key questions and issues that need to be addressed are the following:

1. *House models*. Gudeman (2012: 72–73) observes that his South American ethnography on the house model "throws sparks in multiple directions," demonstrating a "current of vitality" that blurs the division between human society and the environment. Emphasizing the need to seek ways of imagining a better world for humans, non-humans, and the rest of nature, Cronon (1996: 85–86) suggests we pay attention to the "middle ground": "The middle ground is where we actually live. It is where we – all of us, in our different places and ways – make our homes." It now seems pertinent to explore house models – to systematically examine their similarities and difference and their limits and potential for the context of the Anthropocene – through both the ethnographic and archival records. This would be a worthwhile and major anthropological and historical task. Attending to ethnographies and "other" voices and models is a precondition for aligning a variety of actors and agencies at various levels of governance under conditions of complexity and uncertainty.
2. *Solidarity and responsibility*. Given both our relatively new role as geological agents and our awareness of it, it is now *our* primary responsibility to observe, to reflect, and to act. Some recent works on the world of biomedicine and its resources provide a significant step forward on how to proceed. Thus, Prainsack and Buyx (2012) outline the history of the concept of solidarity and related terms, their growing salience in political theory and public discourse, and their usefulness in the context of concerns with biobanks, pandemics, and lifestyle-related diseases. The notion and framework of solidarity necessarily invite key questions about participation and authority. Members of the global anthropocenic household are not an undifferentiated "we" – formed by "the simple addition of homologous magnitudes, much as potatoes in a sack form a sack of potatoes," to paraphrase Marx's description, in his *Eighteenth Brumaire* of 1852, of the "great mass" of the French nation. There are important differences among "us," including those of class, gender, ethnicity, sexuality, and history. Such differences imply that one cannot assume a single self-aware agent addressing the Anthropocene:

   > The fact that the crisis of climate change will be routed through all our "anthropological differences" can only mean that, however anthropogenic the current global warming may be in its origins, there is no corresponding "humanity" that in its oneness can act as a political agent. (Chakrabarty 2012: 14)

   For Chakrabarty, contemporary environmental politics is caught in the tension between the universalizing discourse on the species-being of humans and their collective responsibilities, on the one hand, and, on the other hand, postcolonial politics emphasizing difference and particularity. Resolving or amending such tension remains one of the central tasks on the Anthropocenic agenda.

In addition to our differences, our ambitions at times to act collectively for the benefit of sustainable life are thwarted or blocked by what Benson and Kirsch (2010) call the "politics of resignation" – the manufacture of apathy, cynical distance, the hijacking of critique, and the erosion of public trust. While their analysis is based on the powerful role of capitalist corporations in particular empirical contexts (those of tobacco and mining), it has broad relevance for environmental discourse, the problems of apathy expressed in the American vernacularism of "whatever," an implicit assumption of having no political choices and accepting status quo. Under what conditions, Benson and Kirsch (2010: 460) ask, can the politics of resignation become the basis of transformative politics?

3. *Knowing and being.* Finally, a larger issue remains the broader understanding of "what happens when the human starts to be written into the geological record" (Szerszynski 2012: 168). In geological terms, humans have become a geophysical force, aware, at the same time, of both their role and responsibility. Presumably, this is the rationale behind the expected decision of the International Commission on Stratigraphy to add the category of the "Anthropocene" to the previous list of geological epochs. What does the presumed dual role of subject and object imply? Chakrabarty observes that "we now ... have a mode of existence in which we – collectively and as a geophysical force and in ways we cannot experience ourselves – are 'indifferent' or 'neutral' ... to questions of intrahuman justice. We have *run up against our own limits as it were.*" "Call the mode of being a 'species' or something else," he goes on, "but it has no ontology, it is beyond biology, and it acts as a limit to what we also are in the ontological mode" (Chakrabarty 2012: 14; emphasis added).

Perhaps the predicament of the geologists need not puzzle anthropologists studying the nano-scale of the Anthropocene, the implications of the new genetics, biomics, epigenetics, and similar developments for both *Anthropos* and the study of humans. Have we also run up against *our own* limits? After all, in studying ourselves and the ways in which we refashion bodies and relations we lose an element of objectivity. In a sense, this is anthropocentric anthropology *completely* at home. Any account of human life, however, whatever the scale or context (genomes, cells, bodies, households, the planet), is necessarily situated. The neutral gaze from afar, outside the system observed, is an impossibility, a figment of the modernist imagination. As we have seen, it becomes increasingly difficult to maintain any kind of distinction between observers and the observed, nature and nurture, organism and environment, nano and giga. An integration of the biological and the social along the lines of the biosocial turn should facilitate a broad notion of biopolitics, abandoning narrow conceptualizations of life itself, agency, and responsibility.

Human activities increasingly refashion the structures of bodies and genomes, human and non-human, even life itself; and organisms and species increasingly seem conflated and fuzzy – in Darwin's words "entangled." The environment is literally *embodied* in humans through social practices and, moreover, humans themselves partly *consist of* microbial ecosystems. Given the intimate relations of porous bodies and molecular environments, how should human–environmental relations be conceptualized and refashioned? An important issue now is to capture Darwin's insight and to see where it would lead. To what extent would it be meaningful and useful to extend the notion of Anthropocenic impact to include the human body and "life itself"? Might we benefit from applying the same theoretical and political frameworks to the nano-world of bodies, cells, biomes, and genes and the giga-worlds of the globe? What would be "good" governance in such contexts? How is it possible to align public and private interests? What would count as "private," given our entanglements?

As we have seen, life has become center stage for anthropology and related disciplines. As it travels through epistemic space, it has adapted to and co-evolved with scientific perceptions and practices at several fronts. Some of the important historical developments relate to the Cold War, the arrival of modern genetics and bioinformatics, the quest for outer space, the growth of inter-species sensibilities, and the awareness of anthropocenic change. Yet the disciplinary tensions in anthropology along the nature/society divide are as strong as ever. Strangely enough for the study of a single species, life itself and life as such seem to belong to separate intellectual galaxies.

To explain the rapid upsurge in anthropological interest in the body in the 1980s, Martin (1992) drew upon Lévi-Strauss's argument about the fascination with the disappearing primitive; just as the category of the primitive seemed to fade away or prepare for departure, as a result of radical shifts in geopolitics, it seemed to demand anthropological scrutiny and attention. Similarly, Martin 1992: 121) suggested, the recent interest in the body signified its disappearance, "a dramatic transition in body percept and practice . . . the end of one kind of body and the beginning of another." Could it be that we have also seen the end of one kind of life and the beginning of another? Anthropology, in any case, along with several related fields, seems teaming with life, at least in the sense that the idea of "life" has become the center of attention for many anthropologists.

For me, in closing, the notion of "anthropologies of life" primarily refers to comparative explorations of the *making* of life, however fuzzy and unstable life may be and at whatever scale (from minute genomes to the planetary Gaia), through human thinking and practices, in collaboration with a host of other species and entities. Drawing attention to the *activity* of life rather than its essence, emphasizing the verb "to live," may help to illuminate.

When suggesting that "to live is to radiate; it is to organize the milieu from and around a center of reference," Canguilhem (2008: 113–114) insisted that the study of "the living" should not treat individuality as an object but as an "attribute." The radiating center (functional humans, capable of metabolism, being healthy, and staying alive) is the embedded organism, a node in a meshwork of biosocial relations.

Sources and Acknowledgments

The chapters in the book reproduce or draw upon essays previously published. Permissions are acknowledged from the original publishers. While parts of some of the essays appear in two or more chapters, the key sources of published material are detailed here. Extracts from the sources below are also reproduced with permission.

Chapter 1: Orientations in Anthropology. In J.C. Carrier and D.B. Gewertz (eds.) 2013: *The Handbook of Sociocultural Anthropology*. London: Bloomsbury; Biosocial Relations of Production. *Comparative Studies in Society and History*, 2009, 51(2).

Chapter 2: Decode Me! Anthropology and Personal Genomics. *Current Anthropology*, 2012, 53(S5); Spitting Image. *Anthropology Now*, 2009, 1(3).

Extracts from Family Tree DNA. 2008. http://www.familytreedna.com.

Extracts from World Families Forum. 2008. http://www.worldfamilies.net/forum/index.php.

Chapter 3: Genomic Anthropology: Coming in from the Cold? *Current Anthropology*, 2008, 49(4).

Extract from Goetz, T. 2007. "23AndMe Will Decode Your DNA for $1,000. Welcome to the Age of Genomics," *Wired*, November 17. http://archive.wired.com/medtech/genetics/magazine/15-12/ff_genomics?currentPage=4.

Extract from Hamilton, D.P. 2007. "23andMe: Will the Personal-Genomics Company Need Big Pharma to Make Money?," *VentureBeat*, November 19. http://venturebeat.com/2007/11/19/23andme-will-the-personal-genomics-company-need-big-pharma-to-make-money/.

Chapter 4: LabouringMe, LabouringUs. In B. Prainsack, S. Schicktanz, and G. Werner-Felmayer (eds.) 2014: *Genetics as Social Practice: Transdisciplinary Views on Science and Culture*. London: Ashgate.

Chapter 5: Personal Names: Embodiment, Differentiation, Hierarchy, and Belonging. *Science, Technology and Human Values*, published online January 9, 2014.

Chapter 6: Human Variation: A Biosocial Perspective. In M. Meloni, S. Williams, and P. Martin (eds.) 2015: *Biosocial Matters. Sociological Review* monograph.

Chapter 7: Life at the Border: Nim Chimpsky *et al.* In K. Hastrup (ed.) 2013: *Anthropology and Nature*. London: Routledge.

Chapter 8: Celestial Bodies: Lucy in the Sky. In L. Codignola and K.-U. Schrogl (eds.) 2009: *Humans in Outer Space: Interdisciplinary Odysseys*. New York: Springer.

Chapter 9: Enskilment at Sea. *Man*, 1994, 29(4); "Finding One's Sea Legs": Learning, the Process of Enskilment, and Integrating Fishers and Their Knowledge into Fisheries Science and Management. In B. Neis and L. Felt (eds.) 2000: *Finding Our Sea Legs: Linking Fishery People and Their Knowledge with Science and Management.* St. John's: ISER Press.

Chapter 10: Human–Environmental Relations: Orientalism, Paternalism, and Communalism. In P. Descola and G. Palsson (eds.) 1996: *Nature and Society: Anthropological Perspectives*. London: Routledge.

Chapter 11: Nature and Society in the Age of Postmodernity. In A. Biersack and J. Greenberg (eds.) 2006: *Reimagining Political Ecology*. Durham, NC: Duke University Press.

Extracts from various volumes of *Ægir*.

Chapter 12: Excerpt from W.H. Auden's "Musée des Beaux Arts," copyright © 1940 and renewed 1968 by W.H. Auden; from *W.H. Auden Collected Poems* by W.H. Auden. Used by permission of Random House, an imprint and division of Penguin Random House LLC. All rights reserved.

Chapter 13: Orientations in Anthropology. In J.C. Carrier and D.B. Gewertz (eds.) 2013: *The Handbook of Sociocultural Anthropology*. London: Bloomsbury; Biosocial Relations of Production. *Comparative Studies in Society and History*, 2009, 51(2).

Bibliography

ABC News. 2009. http://abcnews.go.com/blogs/technology/2007/12/the-genome-ofj/. Accessed December 10.

Acheson, J.M. 1977. "Technical Skill and Fishing Success in the Maine Lobster Industry," in H. Lechtman and R. Merrill (eds.), *Material Culture: Styles, Organization, and Dynamics of Technology*. St. Paul, MN: West Publishing Co.

1981. "Anthropology of Fishing," *Annual Review of Anthropology* 10: 275–316.

1988. *The Lobster Gangs of Maine*. Hanover and London: University Press of New England.

Acheson, J.M. and J.A. Wilson. 1996. "Order out of Chaos: The Case for Parametric Fisheries Management," *American Anthropologist* 98: 579–594.

Adkins, L. (2005). "The New Economy, Property and Personhood," *Theory, Culture and Society* 22(1): 111–230.

Aldhous, P. and M. Reilly. 2009. "How My Genome Was Hacked: If a New Scientist Reporter's DNA Is Vulnerable, so Is Yours," *New Scientist* 201: 6–9.

Alia, V. 2006. *Names and Nunavut: Culture and Identity in Arctic Canada*. New York: Berghahn.

Almeling, R. 2007. "Selling Genes, Selling Gender: Egg Agencies, Sperm Banks, and the Medical Market in Genetic Material," *American Sociological Review* 72: 319–340.

Ames, K.M. and H.D.G. Maschner. 1999. *People of the Northwest Coast: Their Archaeology and Prehistory*. London: Thames and Hudson.

Anderson, S.R. 2004. *Doctor Doolitte's Delusion: Animals and the Uniqueness of Human Language*. New Haven, CT: Yale University Press.

Aneesh, A. 2006. *Virtual Migration: The Programming of Globalization*. Durham, NC: Duke University Press.

Appadurai, A. 1986. "Introduction: Commodities and the Politics of Value," in A. Appadurai (ed.), *The Social Life of Things: Commodities in Cultural Perspective*. Cambridge: Cambridge University Press.

Ardener, E. 1989. "Social Anthropology and Population," in M. Chapman (ed.), *The Voice of Prophecy and Other Essays*. Oxford: Berghahn.

Arendt, H. 1958. *The Human Condition*. Chicago: University of Chicago Press.

Arumugam, M., J. Raes, E. Pelletier, D. Le Paslier, T. Yamada, *et al.* 2011. "Enterotypes of the Human Gut Microbiome," *Nature* 473(7346): 174–180.

Auden, W.H. 1940. "Musée des Beaux Arts," in *W.H. Auden Collected Poems*. New York: Random House.

Austin, J.L. 1962. *How to Do Things with Words*. Oxford: Oxford University Press.

Bakhtin, M.M. 1986. *Speech Genres and Other Late Essays*. Trans. V.W. McGee, ed. C. Emerson and M. Holmquist. Austin: University of Texas Press.
Bamford, S. 2007. *Biology Unmoored: Melanesian Reflections on Life and Biotechnology*. Berkeley: University of California Press.
Barnes, B. and J. Dupré. 2008. *Genomes and What to Make of Them*. Chicago: University of Chicago Press.
Barth, F. 1966. *Models of Social Organization*. Occasional Paper No. 23. London: Royal Anthropological Institute of Great Britain and Ireland.
1969. *Ethnic Groups and Boundaries: The Social Organization of Culture Differences*. Oslo: Universitetsforlaget.
Bateson, G. 1972. *Steps to an Ecology of Mind*. London: Granada Publishing.
Bateson, M.C. 1984. *With a Daughter's Eye: A Memoir of Margaret Mead and Gregory Bateson*. New York: Washington Square Press.
Battaglia, D. 1995. "Problematizing the Self: A Thematic Introduction," in D. Battaglia (ed.), *Rhetorics of Self-Making*. Berkeley: University of California Press.
2005. "Introduction," in D. Battaglia (ed.), *E.T. Culture: Anthropology in Outerspaces*. Durham, NC: Duke University Press.
Bellwood, P. 2001. "Early Agriculturalist Population Diasporas? Farming, Language, and Genes," *Annual Review of Anthropology* 30: 181–207.
Bender, J. and D.E. Wellbery (eds.). 1991. *Chronotypes: The Construction of Time*. Stanford, CA: Stanford University Press.
Benezra, A., J. DeStephano, and J.I. Gordon. 2012. "Anthropology of Microbes," *Proceedings of the National Academy of Sciences* 109: 6378–6381.
Benhabib, S. 1992. *Situating the Self: Gender, Community and Postmodernism in Contemporary Ethics*. Cambridge: Polity Press.
Bennett, J.W. 1976. *The Ecological Transition: Cultural Ecology and Human Adaptation*. New York: Pergamon Press.
1993. *Human Ecology as Human Behavior: Essays in Environmental and Developmental Anthropology*. New Brunswick, NJ: Transaction Publishers.
Benson, P. and S. Kirsch. 2010. "Capitalism and the Politics of Resignation," *Current Anthropology* 51(4): 459–486.
Benson, S. 2006. "Injurious Names: Naming, Disavowal, and Recuperation in Contexts of Slavery and Emancipation," in G. vom Bruck and B. Bodenhorn (eds.), *The Anthropology of Names and Naming*. Cambridge: Cambridge University Press.
Benton, T. 1991. "Biology and Social Science: Why the Return of the Repressed Should Be Given a (Cautious) Welcome," *Sociology* 25(1): 1–29.
Berlin, I. 1989. *Against the Current: Essays in the History of Ideas*. Oxford: Clarendon Press.
Biosocial Society, The. 2007. http://www.biosocsoc.org/. Accessed August 28.
Bird-David, N. 1993. "Tribal Metaphorization of Human-Nature Relatedness," in K. Milton (ed.), *Environmentalism: The View from Anthropology*. London: Routledge.
Bjarnason, Th. and Th. Thorlindsson. 1993. "In Defense of a Folk Model: The 'Skipper Effect' in the Icelandic Cod Fishery," *American Anthropologist* 95: 371–394.
Bliss, C. 2012. *Race Decoded*. Stanford, CA: Stanford University Press.
Bloch, M. 1991. "Language, Anthropology and Cognitive Science," *Man* 26: 183–198.
Boas, F. 1911. *The Mind of Primitive Man*. Norwood, MA: The MacMillan Company.
1940. *Race, Language, and Culture*. New York: Macmillan.

Bodenhorn, B. 2006. "Calling Into Being: Naming and Speaking Names on Alaska's North Slope," in G. vom Bruck and B. Bodenhorn (eds.), *The Anthropology of Names and Naming*. Cambridge: Cambridge University Press.
Bodenhorn, B. and G. vom Bruck. 2006. "'Entangled in Histories': An Introduction to the Anthropology of Naming," in G. vom Bruck and B. Bodenhorn (eds.), *The Anthropology of Names and Naming*. Cambridge: Cambridge University Press.
Bolnick, D.A., D. Fullwiley, T. Duster, R.S. Cooper, J.H. Fjimura, *et al.* 2007. "The Science and Business of Genetic Ancestry Testing," *Science* 318(October 19): 399–400.
Bordo, S. 1987. *The Flight to Objectivity: Essays on Cartesianism and Culture*. New York: State University of New York Press.
Bourdieu, P. 1978 [1972]. *Outline of a Theory of Practice*. Trans. R. Nice. Cambridge: Cambridge University Press.
1990. *The Logic of Practice*. Trans. R. Nice. Cambridge: Polity Press.
Brightman, R. 1993. *Grateful Prey: Rock Cree Human–Animal Relationships*. Berkeley: University of California Press.
Burch, E.S. Jr. 2006. *Social Life in Northwest Alaska: The Structure of Iñupiaq Eskimo Nations*. Fairbanks: University of Alaska Press.
Byron, R. 1986. *Sea Change: A Shetland Society, 1970–1979*. St. John's: Institute of Social and Economic Research.
Canetti, E. 1984. *The Numbered: A Play*. Trans. C. Stewart. London: Marion Boyars.
Canguilhem, G. 2008 [1965]. *Knowledge of Life*. New York: Fordham University Press.
Carrier, J. 1998. "Introduction," in J. Carrier and D. Miller (eds.), *Virtualism: A New Political Economy*. Oxford: Berg.
Carsten, J. 2004. *After Kinship*. Cambridge: Cambridge University Press.
Carsten, J. and Hugh-Jones, S. 1995. *About the House: Lévi Strauss and Beyond*. Cambridge: Cambridge University Press.
Cavalli-Sforza, L. 2000. *Genes, Peoples, and Languages*. New York: North Point Press.
Chakrabarty, D. 2012. "Postcolonial Studies and the Challenge of Climate Change," *New Literary History* 43(1): 1–18.
Chomsky, N. 1959. "A Review of B.F. Skinner's *Verbal Behavior*," *Language* 35(1): 26–58.
1986. *Knowledge of Language: Its Nature, Origin, and Use*. New York: Praeger.
Chron.com. 2008. http://blogs.chron.com. Accessed December 19.
Clark, N. 2014. "Geo-politics and the Disaster of the Anthropocene," *The Sociological Review* 62(S1): 19–37.
Clifford, J. 2004. "Looking Several Ways: Anthropology and Native Heritage in Alaska," *Current Anthropology* 45: 5–30.
Clifford, J. and G.E. Marcus (eds.). 1986. *Writing Culture: The Poetics and Politics of Ethnography*. Berkeley: University of California Press.
Coetzee, J.M. 1999. *The Lives of Animals*. London: Profile Books.
Cohen, J. 2010. *Almost Chimpanzee: Searching for What Makes Us Human, in Rainforests, Labs, Sanctuaries, and Zoos*. New York: Times Books.
Collier, S.J. and A. Lakoff. 2005. "On Regimes of Living," in A. Ong and S.J. Collier (eds.), *Global Assemblages: Technology, Politics, and Ethics as Anthropological Problems*. Oxford: Blackwell.

Committee to Review Individual Fishing Quotas. 1999. *Sharing the Fish: Toward a National Policy on Individual Fishing Quotas*. Washington, DC: National Academies Press.
Connerton, P. 1989. *How Societies Remember*. Cambridge: Cambridge University Press.
Connolly, W. 2011. *A World of Becoming*. Durham, NC: Duke University Press.
Cox, K., N.G. Tayles, and H.R. Buckley. 2006. "Forensic Identification of 'Race': The Issues in New Zealand," *Current Anthropology* 47: 869–874.
Coy, M. (ed.). 1989. *Apprenticeship: From Theory to Method and Back Again*. New York: State University of New York Press.
Crawford, M. 2007. "The Importance of Field Research in Anthropological Genetics: Methods, Experiences and Results," in M. Crawford (ed.), *Anthropological Genetics: Theory, Methods and Applications*. Cambridge: Cambridge University Press.
Crichton, M. 2006. *Next*. New York: HarperCollins.
Crist, E. 2004. "Can an Insect Speak? The Case of the Honeybee Dance Language," *Social Studies of Science* 34 (1): 7–43.
2013. "On the Poverty of Our Nomenclature," *Environmental Humanities* 3: 129–147.
Cronon, W. 1996. "The Trouble with Wilderness; Or, Getting Back to the Wrong Nature," in W. Cronon (ed.), *Uncommon Ground: Rethinking the Human Place in Nature*. New York: W.W. Norton and Company.
Crosby, A.W. 1986. *Ecological Imperialism: The Biological Expansion of Europe, 900–1900*. Cambridge: Cambridge University Press.
Dale, G. 2010. *Karl Polanyi: The Limits of the Market*. Cambridge: Polity Press.
Damas, D. (ed.). 1984. *Handbook of North American Indians, vol. 5, Arctic*. Washington, DC: Smithsonian Institution.
Daniau, A.-L., F. d'Errico, and M.F. Sánchez Goñi. 2010. "Testing the Hypothesis of Fire Use for Ecosystem Management by Neanderthal and Upper Palaeolithic Modern Human Populations," *Plos One*. February 11: http://www.plosone.org/article/info%3Adoi%2F10.1371%2Fjournal.pone.0009157.
Darwin, C. 2008 [1859]. *Evolutionary Writings*. Ed. J.A. Secord. Oxford: Oxford University Press.
Davis, D.B. 2006. *Inhuman Bondage: The Rise and Fall of Slavery in the New World*. Oxford: Oxford University Press.
deCODEme. 2009. http://www.decodeme.com/. Accessed March 10.
Derrida, J. 1985. "Des Tours de Babel," in J.F. Graham (ed.), *Difference in Translation*. Ithaca, NY: Cornell University Press.
1994. *Specters of Marx: The State of the Debt, the Work of Mourning, and the New International*. New York: Routledge.
2008. *The Animal that Therefore I Am*. Ed. M.-L. Mallet. Trans. D. Wills. New York: Fordham University Press.
Descola, P. and G. Palsson (eds.). 1996. *Nature and Society: Anthropological Perspectives*. London: Routledge.
Desmond, A. and J. Moore. 2009. *Darwin's Sacred Cause: How a Hatred of Slavery Shaped Darwin's Views on Human Evolution*. Boston, MA: Houghton Mifflin Harcourt.
Dewey, J. 1958 [1929]. *Experience and Nature*. New York: Dover Publications.

Dickenson, D. 2007. *Property in the Body: Feminist Perspectives*. Cambridge: Cambridge University Press.
2013. *Me Medicine vs. We Medicine: Reclaiming Biotechnology for the Common Good*. New York: Columbia University Press.
Dienekes' Anthropology. 2008. http://dienekes.blogspot.com/. Accessed December 8.
Donham, D.L. 1990. *History, Power, Ideology: Central Issues in Marxism and Anthropology*. Cambridge and Paris: Cambridge University Press and Editions de la Maison des Sciences de l'Homme.
Dreyfus, H.L. 1991. *Being-in-the-World: A Commentary on Heidegger's Being and Time, Division I*. Cambridge, MA: The MIT Press.
Duden, B. 1993. *Disembodying Women: Perspectives on Pregnancy and the Unborn*. Cambridge, MA: Harvard University Press.
Dunn, M., S.J. Greenhill, S.C. Levinson, and R.D. Gray. 2011. "Evolved Structure of Language Shows Lineage-Specific Trends in Word-Order Universals," *Nature* 473: 79–82.
Dupré, J. 2010. "The Polygenomic Organism," *The Sociological Review* 58(S1): 19–31.
2012. *Processes of Life: Essays in the Philosophy of Biology*. Oxford: Oxford University Press.
Durkheim, É. 1971 [1912]. *The Elementary Forms of the Religious Life*. London: Allen and Unwin.
Edelman, B. 1993. "Acting Cool and Being Safe: The Definition of Skill in a Swedish Railway Yard," in G. Palsson (ed.), *Beyond Boundaries: Understanding, Translation and Anthropological Discourse*. Oxford: Berg.
Ellen, R. 1982. *Environment, Subsistence and System: The Ecology of Small-Scale Social Formations*. Cambridge: Cambridge University Press.
1996. "Introduction," in R. Ellen and Fukui, K. (eds.), *Redefining Nature: Ecology, Culture and Domestication*. Oxford: Berg.
Engels, F. 2007. "On the Part Played by Labor in the Transition from Ape to Man," in M. Lock and J. Farquardt (eds.), *Beyond the Body Proper*. Durham, NC: Duke University Press.
England, P. 1993. "The Separative Self: Androcentric Bias in Neoclassical Assumptions," in M.A. Ferber and J.A. Nelson (eds.), *Beyond Economic Man: Feminist Theory and Economics*. Chicago and London: University of Chicago Press.
Evans, B. and J. Reid. 2014. *Resilient Life: The Art of Living Dangerously*. Oxford: Polity Press.
Evans, N. and S.C. Levinson. 2009. "The Myth of Language Universals: Language Diversity and Its Importance for Cognitive Science," *Behavioral and Brain Sciences* 32: 429–492.
Family DNA. 2008a. http://www.familydna.co.uk. Accessed December 10.
2008b. http://www.familydna.co.uk. Accessed December 18.
Family Tree DNA. 2008. http://www.familytreedna.com. Accessed December 18.
Fassin, D. 2009. "Another Politics of Life Is Possible," *Theory, Culture & Society* 26(5): 44–60.
Feuchtwang, S. 2006. "Images of Sub-Humanity and Their Realization," *Critique of Anthropology* 26(5): 259–278.
Fienup-Riordan, A. 2000. *Hunting Traditions in a Changing World: Yup'ik Lives in Alaska Today*. Rutgers, NJ: Rutgers University Press.

Finch, J. 2008. "Naming Names: Kinship, Individuality and Personal Names," *Sociology* 42(4): 709–725.

Finney, B.R. and E.M. Jones (eds.). 1985. *Interstellar Migration and the Human Experience*. Berkeley: University of California Press,.

Fischer, K.W., E.J. Rotenberg, D.H. Bullock, and P. Raya. 1993. "The Dynamics of Competence: How Context Contributes Directly to Skill," in R.H. Wozniak and K.W. Fischer (eds.), *Development in Context: Acting and Thinking in Specific Environments*. Hillsdale, NJ: Lawrence Erlbaum Associates.

Fisher, M.S. and G. Downey (eds.). 2006. *Frontiers of Capital: Ethnographic Reflections on the New Economy*. Durham, NC: Duke University Press.

Fleck, L. 1979 [1935]. *Genesis and Development of a Scientific Fact*. Chicago: University of Chicago Press.

Foley, W.A. 2005. "Do Humans Have Innate Mental Structures? Some Arguments from Linguistics," in S. McKinnon and S. Silverman (eds.), *Complexities: Beyond Nature and Nurture*. Chicago: University of Chicago Press.

Forster, P., R. Harding, A. Torroni, and H.J. Bandelt. 1996. "Origin and Evolution of Native American mtDNA Variation: A Reappraisal," *American Journal of Human Genetics* 59: 935–945.

Fortenberry, D. 2013. "The Uses of Race and Ethnicity in Human Microbiome Research," *Trends in Microbiology* 21(4): 165–166.

Fortescue, M. 1997. "Dialect Distribution and Small Group Interaction in Greenlandic Eskimo," in P. McConvell and N. Evans (eds.), *Archaeology and Linguistics*. Oxford: Oxford University Press.

Foucault, M. 1973. *The Birth of the Clinic: An Archaeology of Medical Perception*. Trans. A.M. Sheridan. London: Tavistock.

1988. "Technologies of the Self," in L.H. Martin, H. Gutman, and P.H. Hutton (eds.), *Technologies of the Self*. Amherst, MA: University of Massachusetts Press.

1994. "The Birth of Biopolitics," in P. Rabinow (ed.), *Foucault: Ethics, Subjectivity and Truth*, vol. 1. New York: The New Press.

Fox, R. 1975. "Introduction," in R. Fox (ed.), *Biosocial Anthropology*. London: Malaby Press.

Franklin, S. 2003. "Re-Thinking Nature–Culture: Anthropology and the New Genetics," *Anthropological Theory* 3(1): 65–85.

2013. *Biological Relatives: IVF, Stem Cells, and the Future of Kinship*. Durham, NC: Duke University Press.

Freeman, C. 2011. "Embodying and Affecting Neoliberalism," in F. Mascia-Lees (ed.), *A Companion to the Anthropology of the Body and Embodiment*. Oxford: Wiley-Blackwell.

Friesen, T.M. 2004. "Contemporaneity of Dorset and Thule Cultures in the North American Arctic: New Radiocarbon Dates from Victoria Island, Nunavut," *Current Anthropology* 45: 685–691.

Gare, A.E. 1995. *Postmodernism and the Environmental Crisis*. London: Routledge.

Gatewood, J.B. 1983. "Deciding Where to Fish: The Skippers' Dilemma in Southeast Alaskan Salmon Seining," *Coastal Zone Management Journal* 10: 347–367.

1985. "Actions Speak Louder than Words," in J.W.D. Dougherty (ed.), *Directions in Cognitive Anthropology*. Urbana and Chicago: University of Illinois Press.

Gatewood, J.B. and P.M. Mace. 1990. "Hard Work versus Patience: Simulation of Fishing Strategies in Nova Scotian Purse Seining," *Journal of Quantitative Anthropology* 2(4): 329–352.
Geertz, C. 1973. *The Interpretation of Cultures*. New York: Basic Books.
Geertz, H. and C. Geertz. 1964. "Teknonymy in Bali: Parenthood, Age-Grading and Genealogical Amnesia," *Journal of the Royal Anthropological Institute of Great Britain and Ireland* 94(2): 94–108.
Gibbon, S. and C. Novas. 2007. "Introduction," in S. Gibbon and C. Novas (eds.), *Biosocialities, Genetics and the Social Sciences: Making Biologies and Identities*. London: Routledge.
Gibson, J.J. 1979. *The Ecological Approach to Visual Perception*. Boston, MA: Houghton-Mifflin.
Gibson, K.R. and T. Ingold (eds.). 1993. *Tools, Language and Cognition in Human Evolution*. Cambridge: Cambridge University Press.
Gibson-Graham, J.K. 2011. "A Feminist Project of Belonging for the Anthropocene," *Gender, Place and Culture* 18: 1–21.
Gladwin, T. 1964. "Culture and Logical Process," in W.H. Goodenough (ed.), *Explorations in Cultural Anthropology*. New York: McGraw-Hill.
Glick, T.F. 2008. "The Anthropology of Race across the Darwinian Revolution," in H. Kuklick (ed.), *A New History of Anthropology*. Oxford: Wiley-Blackwell.
Godelier, M. 1986. *The Mental and the Material*. London: Verso.
2009. *In and Out of the West: Reconstructing Anthropology*. London: Verso.
Goetz, T. 2007. "23AndMe Will Decode Your DNA for $1,000. Welcome to the Age of Genomics," *Wired*, November 17. http://archive.wired.com/medtech/genetics/magazine/15-12/ff_genomics?currentPage=4. Accessed May 19, 2015.
Goodall, J. 1971. *In the Shadow of Man*. Boston, MA: Houghton Mifflin.
Goodman, A. 2013. "Bringing Culture into Human Biology and Biology Back into Anthropology," *American Anthropologist* 115(3): 350–373.
Goodwin, C. and A. Duranti. 1992. "Rethinking Context: An Introduction," in A. Duranti and C. Goodwin (eds.), *Rethinking Context: Language as an Interactive Phenomenon*. Cambridge: Cambridge University Press.
Gottweis, H. and A. Peterson (eds.). 2008. *Biobanks: Governance in Comparative Perspective*. London: Routledge.
Gottweis, H. and R. Triendl. 2006. "South Korean Policy Failure and the Hwang Debacle," *Nature Biotechnology* 24(2): 141–143.
Gould, S.J. 1996. *The Mismeasure of Man*. New York: W.W. Norton and Company.
Gravlee, C.C., H.R. Bernard, and W.R. Leonard. 2003. "Heredity, Environment, and Cranial Form: A Reanalysis of Boas's Immigrant Data," *American Anthropologist* 105(1): 125–138.
Griffiths, P.E. and R.D. Gray. 1998. "Developmental Systems and Evolutionary Explanations," in D.L. Hull and M. Ruse (eds.), *The Philosophy of Biology*. Oxford: Oxford University Press.
Grimshaw, J. 1986. *Philosophy and Feminist Thinking*. Minneapolis: University of Minneapolis Press.
Guardian. 2007. http://www.theguardian.com/science/gallery/2007/apr/27/1. Accessed May 19, 2015.
Gudeman, St. 1992. "Remodeling the House of Economics: Culture and Innovation," *American Ethnologist* 19(1): 141–154.
2012. "Vital Energy: The Current of Relations," *Social Analysis* 56(1): 57–73.

Gudeman, S. and A. Rivera. 1990. *Conversations in Colombia: The Domestic Economy in Life and Text*. Cambridge: Cambridge University Press.
1995. "From Car to House (Del coche a la casa)," *American Anthropologist* 97(2): 242–250.
Gurevich, A. 1992. *Historical Anthropology of the Middle Ages*. Cambridge: Polity Press.
Habermas, J. 1990. "Discourse Ethics: Notes on a Program of Philosophical Justification," in S. Benhabib and F. Dallmar (eds.), *The Communicative Ethics Controversy*. Cambridge, MA: The MIT Press.
Hacking, I. 2006. "Genetics, Biosocial Groups and the Future of Identity," *Daedalus* 135(4): 81–95.
Hamilton, A. 2008. "The Retail DNA Test," *Time*, November 3. http://www.time.com/time/specials/packages/article/0,28804,1852747_1854493,00.html.
Hamilton, D.P. 2007. "23andMe: Will the personal-genomics company need Big Pharma to make money?," *VentureBeat*, November 19. http://venturebeat.com/2007/11/19/23andme-will-the-personal-genomics-company-need-big-pharma-to-make-money/. Accessed November 19.
Haraway, D. 1989. *Primate Visions: Gender, Race, and Nature in the World of Modern Science*. London and New York: Routledge.
1996. "Universal Donors in a Vampire Culture: It's All in the Family: Biological Kinship Categories in the Twentieth-Century United States," in W. Cronon (ed.), *Uncommon Ground: Rethinking the Human Place*. New York: W.W. Norton and Company.
2008. *When Species Meet*. Minneapolis: University of Minnesota Press.
Harding, S. 2008. *Sciences from Below: Feminisms, Postcolonialities, and Modernities*. Durham, NC: Duke University Press.
Hardt, M. and A. Negri. 2000. *Empire*. Cambridge, MA: Harvard University Press.
Harré, R. 2009. *Pavlov's Dogs and Schrödinger's Cat: Scenes from the Living Laboratory*. Oxford: Oxford University Press.
Harris, J.F. 2011. *The Ascent of Man: A Philosophy of Human Nature*. London: Transaction Publishers.
Harris, O.J.T. and J. Robb. 2013a. "O Brave New World, That Has Such People In It," in J. Robb and O.J.T. Harris (eds.), *The Body in History: Europe from the Palaeolithic to the Future*. Cambridge: Cambridge University Press.
2013b. "The Body in History: A Concluding Essay," in J. Robb and O.J.T. Harris (eds.), *The Body in History: Europe from the Palaeolithic to the Future*. Cambridge: Cambridge University Press.
Harris, R. 1988. *Language, Saussure and Wittgenstein: How to Play Games with Words*. London: Routledge.
Harvey, D. 1989. *The Condition of Postmodernity*. Oxford: Basil Blackwell.
Hastrup, K. 2007. "Ultima Thule: Anthropology and the Call of the Unknown," *Journal of the Royal Anthropological Institute, n.s.,* 13: 789–804.
(ed.). 2013. *Anthropology and Nature*. London: Routledge.
Hauser, M.D., N. Chomsky, and W.T. Fitch. 2002. "The Faculty of Language: What Is It, Who Has It, and How Did It Evolve?," *Science* 298: 1569–1579.
Heath, A.F. 1976. "Decision Making and Transactional Theory," in B. Kapferer (ed.), *Transaction and Meaning: Directions in Anthropology of Exchange and Symbolic Behavior*. Philadelphia: Institute for the Study of Human Issues.

Heath, D. 1997. "Bodies, Antibodies, and Modest Interventions: Works of Art in the Age of Cyborgian Reproduction," in G. Downey and J. Dumit (eds.), *Cyborgs and Citadels: Anthropological Interventions in the Borderland of Technoscience*. Santa Fe: School of American Research.

Helgason, A. and G. Palsson.1997. "Contested Commodities: The Moral Landscape of Modernist Regimes," *Journal of the Royal Anthropological Institute (incorporating Man)* 3(3): 451–471.

Helgason, A., G. Palsson, H.S. Pedersen, E. Angulalik, E. Dröfn Gunnarsdóttir, *et al.* 2006. "mtDNA Variation in Inuit Populations of Greenland and Canada: Migration History and Population Structure in the Arctic," *American Journal of Physical Anthropology* 130: 123–134.

Hess, C. and E. Ostrom (eds.). 2006. *Understanding Knowledge as a Commons: From Theory to Practice*. Cambridge, MA: The MIT Press.

Hess, E. 2008. *Nim Chimpsky. The Chimp Who Would Be Human*. New York: Bantam Books.

Hoeyer, K. and L. Koch. 2006. "The Ethics of Functional Genomics: Same, Same, but Different?," *Trends in Biotechnology* 25(9): 387–389.

Holland, D. and N. Quinn. 1987. "Culture and Cognition," in D. Holland and N. Quinn (eds.), *Cultural Models in Language and Thought*. Cambridge: Cambridge University Press.

Holling, C.S., L. Gunderson, and G. Peterson. 1994. "Comparing Ecological and Social Systems." Paper presented at the Beijer International Institute of Ecological Economics, The Royal Swedish Academy of Sciences, Stockholm, August 29–30.

Hollingshead, A.B. 1940. "Human Ecology and Human Society," *Ecological Monographs* 10(3): 354–366.

Hooton, E.A. 1946 [1931]. *Up From the Ape*. New York: The Macmillan Company.

Hoyme, L.E. 1953. "Physical Anthropology and Its Instruments: An Historical Study," *Southwestern Journal of Anthropology* 9(4): 408–430.

Hutchins, E. 1983. "Understanding Micronesian Navigation," in D. Genter and A. Stevens (eds.), *Mental Models*. Hillsdale, NJ: Erlbaum.

Ingold, T. 1988. "Introduction," in T. Ingold (ed.), *What Is an Animal?* London: Unwin Hyman.

1993. "Technology, Language, Intelligence: A Reconsideration of Basic Boncepts," in K.R. Gibson and T. Ingold (eds.), *Tools, Language and Cognition in Human Evolution*. Cambridge: Cambridge University Press.

2000. *The Perception of the Environment: Essays on Livelihood, Dwelling and Skill*. London: Routledge.

2001. "From Complementarity to Obviation: On Dissolving the Boundaries Between Social and Biological Anthropology, Archaeology, and Psychology," in S. Oyama, P.E. Griffith, and R.D. Gray (eds.), *Cycles of Contingency: Developmental Systems and Evolution*. Cambridge, MA: The MIT Press.

2007. "Earth, Sky, Wind, and Weather," *Journal of the Royal Anthropological Institute* 13: S19–S38.

2011. *Being Alive: Essays on Movement, Knowledge and Description*. London: Routledge.

2013. "Introduction," in T. Ingold and G. Palsson (eds.), *Biosocial Becomings: Integrating Social and Biological Anthropology*. Cambridge: Cambridge University Press.

Ingold, T. and G. Palsson (eds.). 2013. *Biosocial Becomings: Integrating Social and Biological Anthropology*. Cambridge: Cambridge University Press.
Jablonski, N. 2012. *Living Color: The Biological and Social Meaning of Skin Color*. Berkeley: University of California Press.
Jackson, M. 2005. "What's in a Name?: An Essay on the Power of Words," in *Existential Anthropology*. New York: Berghahn Books
Jasanoff, S. (ed.). 2011. *Reframing Rights:Bioconstitutionalism in the Genetic Age*. Cambridge, MA: The MIT Press.
Jeffrey, I.B., M.J. Claesson, and W. O'Toole. 2012. "Categorization of the Gut Microbiota: Enerotypes or Gradients?," *Nature Reviews* 10(September): 591–592.
Joas, H. 1993. *Pragmatism and Social Theory*. Chicago and London: University of Chicago Press.
Johnson, B. 1985. "Taking Fidelity Philosophically," in J.F. Graham (ed.), *Difference in Translation*. Ithaca, NY: Cornell University Press.
Johnson, T. 1979. "Work Together, Eat Together: Conflict and Conflict Management in a Portuguese Fishing Village," in R. Andersen (ed.), *North Atlantic Maritime Cultures: Anthropological Essays on Changing Adaptations*. The Hague: Mouton Publishers.
Jonas, H. 2001 [1966]. *The Phenomenon of Life*. With a foreword by L. Vogel. Evanston, IL: Northwestern University Press.
Jones, D. 2003. "Kinship and Deep History: Exploring Connections between Culture Areas, Genes, and Languages," *American Anthropologist* 105: 501–514.
Journal of Biosocial Science. (2007). https://www.cambridge.org/journals/journal_catalogue.asp?mnemonic=JBS. Accessed August 28.
Kafka, F. 1983 [1917]. "A Report to the Academy," in N.N. Glatzer (ed.), *The Complete Short Stories of Franz Kafka*. London: Mandarin.
Kahn, P.H. 2011. *Technological Nature: Adaptation and the Future of Human Life*. Cambridge, MA: The MIT Press.
Keita, S.O.Y., R.A. Kittles, C.D.M. Royal, G.E. Bonney, P. Furbert-Harris, *et al.* 2004. "Concepualizing Human Variation," *Nature Genetics Supplement* 36(11): S17–S27.
Keller, E.F. 2000. *The Century of the Gene*. Cambridge, MA: Harvard University Press.
2003. *Making Sense of Life: Explaining Biological Development with Models, Metaphors amd Machines*. Cambridge, MA: Harvard University Press.
Kelly, S.E. 2012. "The Maternal–Foetal Interface and Gestational Chimerism: The Emerging Importance of Chimeric Bodies," *Science as Culture* 1: 1–25. Accessed February 21.
Ketabgian, T. 1996. "The Human Prosthesis: Workers and Machines in the Victorian Industrial Scene," *Critical Matrix* 11: 5–32.
Kirksey, S.E. and S. Helmreich. 2010. "The Emergence of Multispecies Ethnography," *Cultural Anthropology* 25(4): 545–576.
Knoppers, B.M. and R. Chadwick. 2005. "Human Genetic Research: Emerging Trends in Ethics," *Nature Reviews Genetics* 6: 75–79.
Knorr-Cetina, K. 1999. *Epistemic Cultures: How the Sciences Make Knowledge*. Cambridge, MA: Harvard University Press.
Knowlton, D. 1992. "No One Can Serve Two Masters or Native Anthropologist as Oxymoron," *International Journal of Moral and Social Studies* 7: 72–88.
Köhler, W. 1956 [1925]. *The Mentality of Apes*. London: Routledge and Kegan Paul.

Kohn, E. 2007. "How Dogs Dream: Amazonian Natures and the Politics of Transgression," *American Ethnologist* 34(1): 3–24.

Kolbert, E. 2011. "Sleeping with the Enemy: What Happened between the Neanderthals and Us?," *New Yorker*, August 15: 64–75.

2013. "The Lost World: Fossils of the Future," *New Yorker*, December 23.

Kortlandt, A. 1986. "The Use of Stone Tools by Wild-Living Chimpanzees and Earliest Hominids," *Journal of Human Evolution* 15: 77–132.

Kral, M.J. and L. Idlout. 2006. "Participatory Anthropology in Nunavut," in P. Stern and L. Stevenson (eds.), *Critical Inuit Studies: An Anthology of Contemporary Arctic Ethnography*. London: University of Nebraska Press.

Krause, J., C. Lalueza-Fox, L. Orlando, W. Enard, R.E. Green, *et al.* 2007. "The Derived *FOXP2* Variant of Modern Humans Was Shared with Neandertals," *Current Biology* 17: 1908–1912.

Krimsky, S. and J. Gruber (eds.). 2013. *Genetic Explanations: Sense and Nonsense*. Cambridge, MA: Harvard University Press.

Lai, C.S.L., S.E. Fisher, J.A. Hurst, F. Vargha-Khadem, and A.P Monaco. 2001. "A Forkhead-Domain Gene Is Mutated in a Severe Speech and Language Disorder," *Nature* 413: 519–523.

Landecker, H. 2007. *Culturing Life: How Cells Became Technologies*. Cambridge, MA: Harvard University Press.

Landecker, H. and A. Panofsky. 2013. "From Social Structure to Gene Regulation, and Back: A Critical Introduction to Environmental Epigenetics for Sociology," *Annual Review of Sociology* 39: 333–357.

Lasch, C. 1979. *The Culture of Narcissism: American Life in the Age of Diminishing Expectations*. New York: W.W. Norton and Company.

Lassiter, L.E. 2005. *The Chicago Guide to Collaborative Ethnography*. Chicago: University of Chicago Press.

Last, A. 2013. "Negotiating the Inhuman: Bakhtin, Materiality and the Instrumentalization of Climate Change," *Theory, Culture and Society* 30: 60–83.

Latour, B. 2013. "Facing Gaia: A New Enquiry into Natural Religion," The Gifford Lectures, University of Edinburgh, February. http://www.bruno-latour.fr/node/486.

Latour, B. and S. Woolgar. 1979. *Laboratory Life: The Social Construction of Scientific Facts*. Beverly Hills, CA: Sage Publications.

Lave, J. 1988. *Cognition in Practice: Mind, Mathematics and Culture in Everyday Life*. Cambridge: Cambridge University Press.

Lave, J. and E. Wenger. 1991. *Situated Learning: Legitimate Peripheral Participation*. Cambridge: Cambridge University Press.

Le Goff, J. 1988. *The Medieval Imagination*. Chicago: University of Chicago Press.

Leach, E.R. 1970. "The Epistemological Background to Malinowski's Empiricism," in R. Firth (ed.), *Man and Culture: An Evaluation of the Work of Malinowski*. London: Routledge and Kegan Paul.

Lefevere, A. and S. Bassnett. 1990. "Introduction: Proust's Grandmother and the Thousand and One Nights: The 'Cultural' Turn in Translation Studies," in S. Bassnett and A. Lefevere (eds.), *Translation, History and Culture*. London: Pinter Publishers.

Lemke, T. 2014. "New Materialisms: Foucault and the 'Government of Things,'" *Theory, Culture & Society* 0263276413519340, first published on April 2, 2014.

Lever, J. 2008. *Me Cheeta: The Autobiography*. London: Fourth Estate.

Lévi-Strauss, C. and D. Eribon. 1988. *Conversations with Claude Lévi-Strauss.* Chicago: University of Chicago Press.
Levina, M. 2010. "Googling Your Genes: Personal Genomics and the Discourse of Citizen Bioscience in the Network Age," *Journal of Science Communication* 9(1): 1–8.
Levinson, S.C. 2006. "On the Human 'Interaction Engine,'" in N.J. Enfield and S.C. Levinson (eds.), *Roots of Human Sociality: Culture, Cognition and Interaction.* Oxford: Berg.
2009. "Language and Mind: Let's Get the Issues Straight!," in S.D. Blum (ed.), *Making Sense of Language: Readings in Culture and Communication.* Oxford: Oxford University Press.
Lieberman, D.E. 2007. "Homing in On Early Homo," *Nature* (September): 291–292.
Lien, M.E. 2015. *Becoming Salmon: Aquaculture and the Domestication of Fish.* Berkeley: University of California Press.
Lifshitz-Goldberg, Y. 2010. "Gone With the Wind? The Potential Tragedy of the Common Wind," *Journal of Environmental Law and Policy* 28: 435–470.
Lindee, S. and R. Ventura Santos. 2012. "The Biological Anthropology of Living Human Populations: World Histories, National Styles, and International Networks," *Current Anthropology* 53(S5): 1–14.
Lloyd, W. 1977 [1833]. "On the Checks to Population," in G. Hardin and J. Baden (eds.), *Managing the Commons.* San Francisco: W.H. Freeman.
Lock, M. 1993. "Cultivating the Body: Anthropology and Epistemologies of Bodily Practice and Knowledge," *Annual Review of Anthropology* 22: 133–155.
2005. "Eclipse of the Gene and the Return of Divination," *Current Anthropology* 46: S47–S70.
2007. "Human Body Parts as Therapeutic Tools: Contradictory Discourses and Transformed Subjectivities," in M. Lock and J. Farquhar (eds.), *Beyond the Body Proper: Reading the Anthropology of Material Life.* Durham, NC: Duke University Press.
2015. "Comprehending Embodiment in the Era of the Epigenome," *Current Anthropology* 56(2): 151–177.
Lock, M. and P. Kaufert. 2001. "Menopause, Local Biologies and Cultures of Aging," *American Journal of Human Biology* 13(4): 494–504.
Lock, M. and V.-K. Nguyen. 2010. *An Anthropology of Biomedicine.* Oxford: Wiley-Blackwell.
Ludwig, D., R. Hilborn, and C. Walters. 1993. "Uncertainty, Resource Exploitation, and Conservation: Lessons from History," *Science* 260: 17–36.
Luria, A. and L. Vygotsky 1992 [1930]. *Ape, Primitive Man, and Child: Essays in the History of Behavior.* Orlando, FL: CRC Press.
Macauley, D. 1996. "Hanna Arendt and the Politics of Place: From Earth Alienation to Oikos," in D. Macauley (ed.), *Minding Nature: The Philosophers of Ecology.* NewYork: Guilford Press.
Malinowski, B. 1972 [1922]. *Argonauts of the Western Pacific: An Account of Native Enterprise and Adventure in the Archipelagoes of Melanesian New Guinea.* London: Routledge and Kegan Paul.
Marglin, S.A. 1990. "Losing Touch: The Cultural Conditions of Worker Accommodation and Resistance," in F.A. Marglin and S.A. Marglin (eds.), *Dominating Knowledge: Development, Culture and Resistance.* Oxford: Clarendon Press.

Marks, J. 2001. "We're Going to Tell These People Who They Really Are: Science and Relatedness," in S. Franklin and S. McKinnon (eds.), *Relative Values: Reconfiguring Kinship Studies*. Durham, NC: Duke University Press.

2002. *What It Means to Be 98% Chimpanzee: Apes, People, and Their Genes.* Berkeley: University of California Press.

2008. "Race across the Physical–Cultural Divide in American Anthropology," in H. Kuklick (ed.), *A New History of Anthropology*. Oxford: Blackwell.

Martin, A. 2010. "Microchimerism in the Mother(land): Blurring the Borders of Body and Nation," *Body & Society* 16(3): 23–50.

Martin, E. 1987. *The Woman in the Body: A Cultural Analysis of Reproduction*. Boston, MA: Beacon.

1992. "The End of the Body?," *American Ethnologist* 19: 121–140.

Martin, G. 2009. "Spitting image," *The Phrase Finder*. www.phrases.org.uk/meanings/spitting-image.html. Accessed May 11.

Marx, K. 1959 [1884]. *Economic and Philosophical Manuscripts of 1884*. In *Karl Marx and Friedrich Engels: Collected Works*, vol. 3. Moscow: Progress Publishers.

1976 [1867]. *Capital: A Critique of Political Economy*, vol. 1. Trans. Ben Fowkes. Harmondsworth: Penguin Books.

Marx, K. and F. Engels. 1970 [1845]. *The German Ideology*. Edited with an introduction by C.J. Arthur. New York: International Publishers.

Marx, K. with F. Engels. 1998 [1845]. "Theses on Feuerbach," in *The German Ideology*. Amherst, MA: Prometheus Books.

Mary-Rousseliére, G. 1984. "Iglulik," in D. Damas (ed.), *Handbook of North American Indians, vol. 5, Arctic*. Washington, DC: Smithsonian Institution.

Mateos, P. 2007. "A Review of Name-Based Ethnicity Classification Methods and Their Potential in Population Studies," *Populations Space and Place* 13(4): 243–263.

Mauss, M. 1973 [1934]. "Techniques of the Body," *Economy and Society* 2: 70–88.

Maxwell, M.S. 1980. "Archaeology of the Arctic and Subarctic Zones," *Annual Review of Anthropology* 9: 161–185.

McCay, B.J. 1995. "Common and Private Concerns," *Advances in Human Ecology* 4: 89–116.

McCay, B.J. and J.M. Acheson (eds.). 1987. *The Question of the Commons: The Culture and Ecology of Communal Resources*. Tucson: University of Arizona Press.

McGhee, R. 2006. *The Last Imaginary Place: A Human History of the Arctic World*. Oxford: Oxford University Press.

M'charek, A. 2005. *The Human Genome Diversity Project: An Ethnography of Scientific Practice*. Cambridge: Cambridge University Press.

Meloni, M. 2014. "How Biology Became Social, and What It Means for Social Theory," *The Sociological Review* 62(3): 593–614.

Merchant, C. 1990. "Environmental Ethics and Political Conflict: A View from California," *Environmental Ethics* 12(1): 45–68.

Michaels, M. 1999. "Fetal Galaxies: Some Questions about What We See," in L.M. Morgan and M.W. Michaels (eds.), *Fetal Subjects, Feminist Positions*. Philadelphia: University of Pennsylvania Press.

Miller, G.A., E. Galanter, and K.H. Pribram. 1960. *Plans and the Structure of Behavior*. New York: Holt, Rinehart and Winston.

Moran, E. 2006. *People and Nature: An Introduction to Human Ecological Relations*. Oxford: Blackwell.

Morris, R.C. 2007. "Legacies of Derrida: Anthropology," *Annual Review of Anthropology* 36: 355–389.

Moseley, W. 2002. *Futureland*. New York: Aspect.

Moss, L. 2003. *What Genes Can't Do*. Cambridge, MA: The MIT Press.

Nature. 2000. October: 407.

Neild, E. 1989. "Translation Is a Two-Way Street: A Response to Steiner," *Meta* 34(2): 238–241.

Neimanis, A. 2013. "Morning Sickness and Gut Sociality: Towards a Posthumanist Feminist Phenomenology," *Janus Head* 13: 214–240.

Nelson, J.A. 1992. "Gender, Metaphor, and the Definition of Economics," *Economics and Philosophy* 8: 103–125.

1993. "The Study of Choice or the Study of Provisioning? Gender and the Definition of Economics," in M.A. Ferber and J.A. Nelson (eds.), *Beyond Economic Man: Feminist Theory and Economics*. Chicago: University of Chicago Press.

Niewöhner, J. 2011. "Epigenetics: Embedded Bodies and the Molecularisation of Biography and Milieu," *BioSocieties* 6: 279–298.

Nowotny, H., P. Scott, and M. Gibbon. 2001. *Rethinking Science: Knowledge in an Age of Uncertainty*. Cambridge: Polity.

Nuttall, M. 1992. *Arctic Homeland: Kinship, Community, and Development in Northwest Greenland*. Toronto: Toronto University Press.

Ochs, E. 1988. *Culture and Language Development: Language Acquisition and Language Socialization in a Samoan Village*. Cambridge: Cambridge University Press.

Odling-Smee, F.J. 1994. "Niche Construction, Evolution and Culture," in T. Ingold (ed.), *Encyclopedia of Anthropology*. London: Routledge.

Okely, J. 1992. "Anthropology and Autobiography: Participatory Experience and Embodied Knowledge," in J. Okely and H. Callaway (eds.), *Anthropology and Autobiography*. London: Routledge.

Ortner, S.B. 1984. "Theory in Anthropology since the Sixties," *Comparative Studies in Society and History*, 26(1): 126–186.

Ostrom, E. 1990. *Governing the Commons: The Evolution of Institutions for Collective Action*. Cambridge: Cambridge University Press.

2005. *Understanding Institutional Diversity*. Princeton, NJ: Princeton University Press.

Oyama, S. 2000. *The Ontogeny of Information: Developmental Systems and Evolution*. Durham, NC: Duke University Press.

Pagli, C. and F. Sigmundsson. 2008. "Will Present Day Glacier Retreat Increase Volcanic Activity? Stress Induced by Recent Glacier Retreat and Its Effect on Magmatism at the Vatnajökull Ice Cap, Iceland," *Geophysical Research Letters*, 35: 1–5.

Palmié, S. 2007. "Genomics, Divination, 'Racecraft,'" *American Ethnologist* 34: 205–222.

Palsson, G. 1991. *Coastal Economies, Cultural Accounts: Human Ecology and Icelandic Discourse*. Manchester: Manchester University Press.

2001. *Writing on Ice: The Ethnographic Notebooks of Vilhjalmur Stefansson*. Hanover: University Press of New England.

2005. *Travelling Passions: The Hidden Life of Vilhjálmur Stefansson*. Trans. K. Kunz. Winnipeg: University of Manitoba Press and University Press of New England.
2007. *Anthropology and the New Genetics*. Cambridge: Cambridge University Press.
2013. "Ensembles of Biosocial Relations," in T. Ingold and G. Palsson (eds.), *Biosocial Becomings: Integrating Social and Biological Anthropology*. Cambridge: Cambridge University Press.
2016. *The Man Who Stole Himself*. Chicago: University of Chicago Press. (In press).
Palsson, G. and E.P. Durrenberger. 1982. "To Dream of Fish: The Causes of Icelandic Skippers' Fishing Success," *Journal of Anthropological Research* 38: 227–242.
1983. "Icelandic Foremen and Skippers: The Structure and Evolution of a Folk Model," *American Ethnologist* 10: 511–528.
Palsson, G. and S. Örn Guðbjörnsson. 2011. "Knochenarbeit: Die Erfindung des Homo Islandicus," in G. Krüger and M. Sommer (eds.), *Biohistorische Anthropology: Knochen, Körper und DNA in Erinnerungskulturen*. Berlin: Kadmos.
Palsson, G. and A. Helgason. 1995. "Figuring Fish and Measuring Men: The Individual Transferable Quota System in the Icelandic Cod Fishery," *Ocean and Coastal Management* 28(1–3): 117–146.
Palsson, G. and B. Prainsack. 2011. "Genomic Stuff: Governing the (Im)matter of Life," *International Journal of the Commons* 5: 259–283.
Palsson, G., B. Szerszynski, S. Sörlin, J. Marks, B. Avril, *et al.* 2013. "Reconceptualizing the 'Anthropos' in the Anthropocene: Integrating the Social Sciences and Humanities in Global Environmental Change Research," *Environmental Science and Policy* 28: 3–13.
Pálsson, J.Ó.P. 1976. "Rassengeschichte Islands," in I. Schwidetsky (ed.), *Rassengeschichte der Menschheit,* vol. 4. Munich: Oldenbourg.
Panofsky, E. 1991 [1927]. *Perspective as Symbolic Form*. Trans. C.S. Wood, Cambridge, MA: Zone Books.
Patterson, T.C. 2009. *Karl Marx, Anthropologist*. Oxford: Berg.
Pina-Cabral, J. de. 2011. "The Truth about Personal Names," *Journal of the Royal Anthropological Institute* 16: 297–312.
Pinker, S. 1994. *The Language Instinct*. New York: W.W. Norton and Company.
Pitrou, P. 2014. "Life as a Process of Making in the Mixe Highlands (Oaxaca, Mexico): Towards a 'General Pragmatics' of Life," *Journal of the Royal Anthropological Institute* 21(1): 86–105.
Plumwood, V. 1991. "Nature, Self, and Gender: Feminism, Environmental Philosophy, and the Critique of Rationalism," *Hypatia* 6(1): 3–27.
Polanyi, K. 1967. "Aristotle Discovers the Economy," in K. Polanyi, C.M. Arensberg, and H.W. Pearson (eds.), *Trade and Markets in the Early Empires*. Glencoe, IL: Free Press.
Polanyi, M. 1958. *Personal Knowledge: Towards a Post-Critical Philosophy*. Chicago: University of Chicago Press.
Praet, I. 2013. "Humanity and Life as the Perpetual Maintenance of Specific Efforts: A Repraisal of Animism," in T. Ingold and G. Palsson (eds.), *Biosocial Becomings: Integrating Social and Biological Anthropology*. Cambridge: Cambridge University Press.

Prainsack, B. 2011. "Voting with Their Mice: Personal Genome Testing and the 'Participatory Turn' in Disease Research," *Accountability in Research: Policies and Quality Assurance*, 18(3): 132–147.

Prainsack, B. and A. Buyx. 2012. "Solidarity in Contemporary Bioethics: Towards a New Approach," *Bioethics* 26: 343–350.

Prainsack, B., J. Reardon, R. Hindmarsh, H. Gottweis, U. Naue, *et al.* 2008. "Misdirected Precaution," *Nature* 456(6): 34–35.

Prainsack, B., S. Schicktanz, and G. Werner-Felmayer (eds.). 2014. *Genetics as Social Practice: Transdisciplinary Views on Science and Culture*. Farnham: Ashgate.

Preuß, D. 2009. *"Anthropologe und Forschungsreisender": Biographie und Anthropologie Egon Freiherr von Eickstedts (1892–1965)*. Munich: Utz.

Probyn, E. 2013. "Entanglements: Fish, Guts, and Bio-Cultural Sustainability," in E.-J. Abbots and A. Lavis (eds.), *Why We Eat, How We Eat: Contemporary Encounters between Foods and Bodies*. London: Ashgate.

Proctor, R. 1988. "From Anthropologie to Rassenkunde in the German Anthropological Tradition," in G.W. Stocking Jr. (ed.), *Bones, Bodies, Behavior: Essays on Biological Anthropology*. Madison: University of Wisconsin Press.

Rabinow, P. 1996. "Artificiality and Enlightenment: From Sociobiology to Biosociality," in *Essays on the Anthropology of Reason*. Princeton, NJ: Princeton University Press.

2007. "Afterword: Concept Work," in S. Gibbon and C. Novas (eds.), *Genetics and the Social Sciences: Making Biosociality*. London: Routledge.

2008. *Marking Time: On the Anthropology of the Contemporary*. Princeton, NJ: Princeton University Press.

Rabinow, P. and G. Bennett. 2012. *Designing Human Practices: An Experiment with Synthetic Biology*. Chicago: University of Chicago Press.

Rapp, R., D. Heath, and K.-S. Taussig. 2001. "Genealogical Dis-ease: Where Hereditary Abnormality, Biomedical Explanation, and Family Responsibility Meet," in S. Franklin and S. McKinnon (eds.), *Relative Values: Reconfiguring Kinship Studies*. Durham, NC: Duke University Press.

Rappaport, R. 1967. *Pigs for the Ancestors: Ritual in the Ecology of a New Guinea People*. New Haven, CT: Yale University Press.

Reardon, J. 2005. *Race to the Finish: Identity and Governance in an Age of Genomics*. Princeton, NJ: Princeton University Press.

Reed, E.S. 1993. "The Intention to Use a Specific Affordance: A Conceptual Framework for Psychology," in R.H. Wozniak and K.W. Fischer (eds.), *Development in Context: Acting and Thinking in Specific Environments*. Hillsdale, NJ: Lawrence Erlbaum Associates.

Rendell, L. and H. Whitehead 2001. "Culture in Whales and Dolphins," *Behavioral and Brain Sciences* 24: 309–382.

Restrepo, E. and A. Escobar. 2005. "Other Anthropologies and Anthropologies Otherwise: Steps to a World Anthropology Framework," *Critique of Anthropology* 25(2): 99–129.

Rheinberger, H.-J. 2000a [1995/1996]. "Beyond Nature and Culture: Modes of Reasoning in the Age of Molecular Biology and Medicine," in M. Lock, A. Young, and A. Cambrosio (eds.), *Living and Working with the New Medical Technologies: Intersections of Inquiry*. Cambridge: Cambridge University Press.

2000b. "Gene Concepts," in P. Beurton, R. Falk, and H.-J. Rheinberger (eds.), *The Concept of the Gene in Development and Evolution*. Cambridge: Cambridge University Press.
2009. "Molecular Biology: A Paradigm of 20th Century Science: On the Nature of Disciplines and on Cultures of Emergence." Paper presented at Mapping Interfaces: The Future of Knowledge, Reykjavik, Iceland, June 16–17.
2010. *An Epistemology of the Concrete: Twentieth-Century Histories of Life*. Durham, NC: Duke University Press.
Richardson, S.S. 2013. *Sex Itself: The Search for Male and Female in the Human Genome*. Chicago: University of Chicago Press.
Richerson, P.J. and R. Boyd. 2008. *Not by Genes Alone. How Culture Transformed Human Evolution*. Chicago: University of Chicago Press.
Ritzer, G. and N. Jurgenson 2010. "Production, Consumption, Prosumption: The Nature of Capitalism in the Age of the Digital 'Prosumer,'" *Journal of Consumer Culture* 10(1): 13–36.
Roberts, D.F. 1970. "Review of *Journal of Biosocial Science*, Vol. I:I," *Man (Journal of the Royal Anthropological Institute)* 5(1): 133–134.
Rockhill, E. 1999. *Genetic Structure and Affinities among Indigenous Populations of Northeast Siberia: A Literature Review*. M.Phil. thesis, Darwin College, Cambridge University.
Rockström, J., W. Steffen, K. Noone, Å. Persson, F.S. Chapin, *et al.* 2009. "A Safe Operating Space for Humanity," *Nature* 461: 472–475.
RootsTelevision. 2008. http://rootstelevision.com/blogs. Accessed December 19.
Rosaldo, R. 1980. *Ilongot Head Hunting 1883–1974: A Study in Society and History*. Stanford, CA: Stanford University Press.
1986. "From the Door of His Tent: The Fieldworker and the Inquisitor," in J. Clifford and G.E. Marcus (eds.), *Writing Culture: The Poetics and Politics of Ethnography*. Berkeley: University of California Press.
1987. "Where Objectivity Lies: The Rhetoric of Anthropology," in J.S. Nelson, A. Megill, and D.N. McCloskey (eds.), *The Rhetorics of the Human Sciences*. Madison: University of Wisconsin Press.
Rose, N. 2005. *The Politics of Life Itself: Biomedicine, Power, and Subjectivity in the Twenty-First Century*. Princeton, NJ: Princeton University Press.
Roth, C.F. 2005. "Ufology as Anthropology: Race, Extraterrestrials, and the Occult," in D. Battaglia (ed.), *E.T. Culture: Anthropology in Outerspaces*. Durham, NC: Duke University Press.
2008. *Becoming Tsimshian: The Social Life of Names*. Seattle: University of Washington Press.
Rudwick, M.J.S. 2014. *Earth's Deep History: How It Was Discovered and Why It Matters*. Chicago: University of Chicago Press.
Russell, E. 2004. "The Garden in the Machine: Toward an Evolutionary History of Technology," in S.R. Schrepfer and P. Scranton (eds.), *Industrializing Organisms: Introducing Evolutionary History*. New York: Routledge.
Sahlins, M. 1972. *Stone Age Economics*. London: Tavistock.
1976. *Culture and Practical Reason*. Chicago: University of Chicago Press.
Saladin d'Anglure, B. 1977. "Iqallijuq ou les reminiscences d'une âme-nom inuit," *Études Inuit* 1: 33–63.
Salkin, A. 2008. "When in Doubt, Spit It Out," *New York Times*, September 12.

Santos, R.V., P.H. Fry, S. Monteiro, M.C. Maio, J.C. Rodrigues, *et al.* 2009. "Color, Race, and Genomic Ancestry in Brazil: Dialogues between Anthropology and Genetics," *Current Anthropology* 50(6): 787–819.

Saussure, F. de 1959 [1916]. *Course in General Linguistics*. New York: McGraw-Hill.

Savage-Rumbaugh, S. and R. Lewin 1994. *Kanzi: The Ape at the Brink of the Human Mind*. New York: Wiley.

Schabas, M. 2005. *The Natural Origins of Economics*. Chicago: University of Chicago Press.

Schmelz, M., J. Call, and M. Tomasello 2011. "Chimpanzees Know that Others Make Inferences," *Proceedings of the National Academy of Sciences of the United States* 108(7): 3077–3079.

Schmidt, A. 1973. *The Concept of Nature in Marx*. London: NLB.

Schneider, I. 2003. "Gesellschaftliche Umgangsweisen mit Keimzellen: Regulation zwischen Gabe, Verkauf und Unveräußerlichkeit," in S. Graumann and I. Schneider (eds.), *Verkörperte Technik – Entkörperte Frau: Biopolitik und Geschlecht*. Frankfurt am Main and New York: Campus.

Schopka, S. 1975. "Fiskispár" ("Fishing Forecast"), *Víkingur*, pp. 47–49. Reykjavík.

Schrödinger, E. 2006. *What Is Life?* Cambridge: Cambridge University Press.

Schurr, T.G. 2004. "The Peopling of the New World: Perspectives from Molecular Anthropology," *Annual Reviews in Anthropology* 33: 551–583.

ScienceBlogs. 2009. http://scienceblogs.com/geneticfuture/. June 2009. Accessed May 5, 2011.

Scott, G.R., S. Legge, R.W. Lane, S.L. Steen, and S.R. Street. 2000. "Physical Anthropology of the Arctic," in M. Nuttall and T.V. Callaghan (eds.), *The Arctic: Environment, People, Policy*. Amsterdam: Harwood Academic Publishers.

Scott, J.C. 1998. *Seeing Like a State: How Certain Schemes to Improve the Human Condition Have Failed*. New Haven, CT: Yale University Press.

Scott, J.C., J. Tehranian, and J. Mathias. 2002. "The Production of Legal Identities Proper to States: The Case of the Permanent Family Surname," *Comparative Studies in Society and History* 44(1): 4–44.

Segal, D.A. and S.J. Yanagisako. 2005. "Introduction," in D.A. Segal and S.J. Yanagisako (eds.), *Unwrapping the Sacred Bundle: Reflections on the Disciplining of Anthropology*. Durham, NC: Duke University Press.

Serre, D. and S. Pääbo. 2004. "Evidence for Gradients of Human Genetic Diversity within and among Continents," *Genome Research* 14: 1679–1685.

Shoreman-Ouimet, E. and Kopina, H. 2011. "Introduction: Environmental Anthropology of Yesterday and Today," in H. Kopina and E. Shoreman-Ouimet (eds.), *Environmental Anthropology*. London: Routledge.

Showstack, R. 2013. "Scientists Debate Whether the Anthropocene Should Be a New Geological Epoch," *Eos* 94: 41–42.

Singer, P. 2011. "The Troubled Life of Nim Chimpsky." *NYRblog*, August 18. http://www.nybooks.com/blogs/nyrblog/2011/aug/18.

Smocovitis, V.B. 2012. "Humanizing Evolution: Anthropology, the Evolutionary Synthesis, and the Prehistory of Biological Anthropology, 1927–1962," *Current Anthropology* 53(suppl. 5).

Snow, C.P. 2001 [1959]. *The Two Cultures*. Cambridge: Canto.

Sofia, Z. 1984. "Exterminating Fetuses: Abortion, Disarmament, and the Sexo-Semiotics of Extraterrestrialism," *Diacritics* 14(2): 47–59.

Sommer, M. 2008. "History in the Gene: Negotiations between Molecular and Organismal Anthropology," *Journal of the History of Biology* 41: 473–528.
Sörlin, S. 2013. "Reconfiguring Environmental Expertise," *Environmental Science and Policy* 28: 14–24.
Spittoon. 2009. http://spittoon.23andme.com/. Accessed May 29.
Steensby, H.P. 1917. "The Eskimo Culture and Theories on Its Origin," *Meddelelser om Grønland* 53: 41–75.
Steffen, W., J. Grinevald, P. Crutzen, and J. McNeill, 2011. "The Anthropocene: Conceptual and Historical Perspectives," *Philosophical Transactions of the Royal Society* 369: 842–867.
Stein, P.L. and B.M. Rowe. 2000. *Physical Anthropology*. Seventh edition. Boston, MA: McGraw-Hill.
Stern, P. and L. Stevenson (eds.). 2006. *Critical Inuit Studies: An Anthology of Contemporary Arctic Ethnography*. London: University of Nebraska Press.
Stevenson, L. 2014. *Life beside Itself: Imaging Care in the Canadian Arctic*. Berkeley: University of California Press.
Stone, L. and P.F. Lurquin. 2005. *A Genetic and Cultural Odyssey: The Life and Work of L. Luca Cavalli-Sforza*. New York: Columbia University Press.
Stone, L., P.F. Lurquin, and L. Luca Cavalli-Sforza. 2007. *Genes, Culture, and Human Evolution: A Synthesis*. Malden, MA: Blackwell Publishing.
Strathern, M. 1980. "No Nature, No Culture: The Hagen Case," in C. MacCormack and M. Strathern (eds.), *Nature, Culture and Gender*. Cambridge: Cambridge University Press.
1990. *Reproducing the Future: Anthropology, Kinship, and the New Reproductive Technologies*. Manchester: University of Manchester Press.
1996. "Cutting the Network," *Journal of the Royal Anthropological Institute* 2(3): 517–528.
2004. "The Whole Person and Its Artifacts," *Annual Review of Anthropology* 33: 1–19.
Suchman, L. 1987. *Plans and Situated Actions: The Problem of Human–Machine Communication*. Cambridge: Cambridge University Press.
Sunder Rajan, K. 2006. *Biocapital: The Constitution of Postgenomic Life*. Durham, NC: Duke University Press.
Sykes, B. 2001. *The Seven Daughters of Eve*. London: Bantam.
Szerszynski, B. 2012. "The End of the End of Nature: The Anthropocene and the Fate of the Human," *The Oxford Literary Review* 34: 165–184.
Tapper, R. 1988. "Animality, Humanity, Morality, Society," in T. Ingold (ed.), *What Is an Animal?* London: Unwin Hyman.
Terrace, H.S. 1987. *Nim: A Chimpanzee Who Learned Sign Language*. New York: Columbia University Press.
2011. "Can Chimps Converse? An Exchange." *New York Review of Books*. http://www.nybooks.com /articles/archives/2011/nov/24/can-chimps-converse-exchange/.
Terrace, H.S., L.A. Petitto, R.J. Sanders, and T.G. Bever 1979. "Can an Ape Create a Sentence?," *Science* 206(4421): 891–902.
Tett, G. 2009. *Fool's Gold: How Unrestrained Greed Corrupted a Dream, Shattered Global Markets and Unleashed a Catastrophe*. London: Little, Brown.
Thacker, E. 2005. *The Global Genome: Biotechnology, Politics, and Culture*. Cambridge, MA: The MIT Press.

Thompson, C. 2005. *Making Parents: The Ontological Choreography of Reproductive Technologies*. Cambridge, MA: The MIT Press.

Thorlindsson, Th. 1988. "The Skipper Effect in the Icelandic Herring Fishery," *Human Organization* 47: 199–212.

Tickell, C. 2011. "Societal Responses to the Anthropocene," *Philosophical Transactions of the Royal Society* 369: 926–932.

Tomasello, M. 2006. "Why Don't Apes Point?," in N.J. Enfield and S.C. Levinson (eds.), *Roots of Human Sociality: Culture, Cognition and Interaction*. Oxford: Berg.

United States District Court for the Northern District of California San Francisco. ANGEL FRALEY, et al. Plaintiffs, v. FACEBOOK, INC., Defendant. 2012. No. C 11-1726 RS. District Judge Richard Seeborg. August 17.

Valentine, D., V.A. Olson, and D. Battaglia (eds.). 2012. "Extreme: Limits and Horizons in the Once and Future Cosmos," *Anthropological Quarterly* 85(4): 1007–1026.

Vayda, A.P. and B.J. McCay. 1975. "New Directions in Ecology and Ecological Anthropology," *Annual Review of Anthropology* 4: 293–306.

Verdery, C. 1999. *The Political Lives of Dead Bodies: Reburial and Postsocialist Change*. New York: Columbia University Press.

Vilaça, A. 2005. "Chronically Unstable Bodies: Reflections on Amazonian Corporealities," *Journal of the Royal Anthropological Institute* 11: 445–464.

Viveiros de Castro, E. 1998. "Cosmological Deixis and Amerindian Perspectivism," *Journal of the Royal Anthropological Institute* 4: 469–488.

vom Bruck, G. and B. Bodenhorn (eds.). 2006. *The Anthropology of Names and Naming*. Cambridge: Cambridge University Press.

Wagner, R. 1977. "Scientific and Indigenous Papuan Conceptualizations of the Innate: A Semiotic Critique of the Ecological Perspective," in T.P. Bayliss-Smith and R.G. Fechem (eds.), *Subsistence and Survival: Rural Ecology in the Pacific*. London: Academic Press.

Wengle, J.L. 1988. *Ethnographers in the Field: The Psychology of Research*. Tuscaloosa and London: University of Alabama Press.

White, D.R.M. 1989. "Knocking 'Em Dead: Alabama Shrimp Boats and the 'Fleet Effect,'" in J.S. Thomas, L. Maril, and E.P. Durrenberger (eds.), *Marine Resource Utilization: A Conference on Social Science Issues*. Mobile: University of South Alabama College of Arts and Sciences Publication, Vol. 1, and the Mississippi-Alabama Sea Grant Consortium.

Williams, R. 1976. *Keywords: A Vocabulary of Culture and Society*. Glasgow: Fontana.

Wilson, J.A. 1990. "Fishing for Knowledge," *Land Economics* 66: 12–29.

World Families Forum. 2008. http://www.worldfamilies.net/forum/index.php. Accessed December 10.

Ægir. 1921. 14. The Journal of the Icelandic Fisheries Association, Reykjavik.
1924. 17. The Journal of the Icelandic Fisheries Association, Reykjavik.
1928. 21. The Journal of the Icelandic Fisheries Association, Reykjavik.
1931. 24. The Journal of the Icelandic Fisheries Association, Reykjavik.
1932. 25. The Journal of the Icelandic Fisheries Association, Reykjavik.
1947. 40. The Journal of the Icelandic Fisheries Association, Reykjavik.
1964. 57. The Journal of the Icelandic Fisheries Association, Reykjavik.
1979. 72. The Journal of the Icelandic Fisheries Association, Reykjavik.
1986. 79. The Journal of the Icelandic Fisheries Association, Reykjavik.

Index

Lightning Source UK Ltd.
Milton Keynes UK
UKOW06n2017071215

264292UK00005B/75/P

9 781107 085848